MICROARRAY GENE EXPRESSION
DATA ANALYSIS

A Beginner's Guide

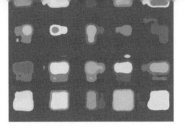

Microarray Gene Expression Data Analysis

A Beginner's Guide

Helen C. Causton
MRC Clinical Sciences Centre
Imperial College
London
UK

John Quackenbush
The Institute for Genome Research
Rockville
Maryland
USA

AND

Alvis Brazma
European Bioinformatics Institute
EMBL Outstation
Wellcome Trust Genome Campus
Hinxton
Cambridge
UK

Blackwell
Publishing

First published 2003
Reprinted 2003

Library of Congress Cataloging-in-Publication Data

Causton, Helen C.
Microarray/gene expressions data analysis: a beginner's guide/Helen C. Causton,
John Quackenbush, and Alvis Brazma.
p. cm.
Includes bibliographical references and index.
ISBN 1-4051-0682-4 (pbk. : alk. paper)
1. DNA microarrays. 2. Gene expression – Research – Methodology. I. Quackenbush,
John. II. Brazma, Alvis. III. Title.
QP624.5.D726   C38   2003
572.8633 – dc21   2002034224

A catalogue record for this title is available from the British Library.

Set in 10/12 pt Galliard
by SNP Best-set Typesetter Ltd, Hong Kong
Printed and bound in the United Kingdom
by MPG Books Ltd, Bodmin, Cornwall

For further information on
Blackwell Publishing, visit our website:
http://www.blackwellpublishing.com

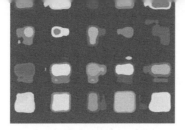

Contents

Chapter 3 Image processing, normalisation and data transformation

Chapter 4 Analysis of gene expression data matrices

Colour plates fall between pp. 84 and 85.

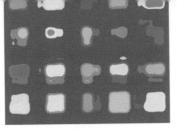

Preface

Microarrays are a tool for monitoring gene expression levels for thousands of genes in parallel. This technology has great utility, as patterns in the gene expression can be used for molecular characterisation of a wide range of diverse phenomena, from disease states and responses to stimuli, to the underlying differences between cells of different types, and for elucidating gene function. Microarray technology is increasingly finding its way into fields as diverse as clinical diagnosis, drug screening and environmental testing.

The amount of information obtained from each microarray experiment can be considerable. This presents new challenges for data storage, management and analysis by life scientists who have not traditionally had to deal with this type or quantity of data. An additional challenge is presented by software for data analysis that has largely been written by statisticians and mathematicians, employing terminology and concepts often unfamiliar to the life scientist.

This book addresses some of the issues faced by researchers in carrying out their first microarray experiments, and covers aspects of designing and analysing the results of microarray experiments. Although microarrays can be used to study phenomena other than gene expression, this remains the most common use of the technology and is the only use of arrays discussed here. The book is not intended to replace the bioinformatician, or statistician, but rather to explain the underlying concepts and principles routinely used in analysis of gene expression data, facilitating sound experimental design, use of the available software and communication with statisticians. The book is also intended for use by statisticians, computer scientists and students of bioinformatics who want a grounding in the types of analysis currently used to study microarray data.

Although all the authors contributed to all sections of the book, Chapter 2 was primarily written by Helen Causton, Chapter 3 by John Quackenbush and Chapter 4 by Alvis Brazma.

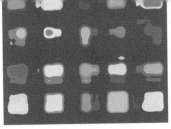

Acknowledgements

Many people have been involved in the production of this book. The authors would like to thank: Tim Aitman, Christiane Albrecht, Peter Green, Inge Jonassen, Norman H. Lee, Fran Lewitter, Clare Marshall, Helen Parkinson, Sylvia Richardson, Ugis Sarkans, Thomas Schlitt, Lev Soinov, Alex Tipping and Jaak Vilo, for valuable comments on the manuscript; and Priti Hegde, Misha Kapushesky, Johan Rung, Alexander Saeed, Vasily Sharov and Ivana Yang, who contributed figures.

Numerous others contributed to the book, either directly or indirectly, including many members of the MGED Board of Directors, Helen Banks, John Barnett, Peter Broderick, Tom Freeman, Michelle Gaasenbeek, Chris Higgins, Ezra Jennings, Mahendra Navarange, Joan Richmond, Ann Schlesinger, Martin Vingron, Peter Young and Rick Young.

Last, but by no means least, the authors would also like to thank their families, who patiently coped with many lost weekends and holidays. In particular they thank Diana Brazma, Mary Kalamaras and Ben Liebman, who put up with even more distant and distracted partners than usual.

The authors would like to thank Marta De Menezes-Graca, Artist in Residence at the MRC Clinical Sciences Centre (2001–2002), for the image used at the beginning of each chapter.

CHAPTER 1

Introduction

Knowledge is the process of piling up facts; wisdom lies in their simplification.

Martin H. Fischer

The last 5–10 years have brought spectacular achievements in genome sequencing – five higher organisms and over 60 microbes have been sequenced, the draft human genome has been published, and substantial parts of several other eukaryotic genomes are now known. However, genome sequencing is nothing more than the transfer of information from one digital carrier – DNA – to another – the electronic computer. Even if we assume that all the genes have been correctly identified, the result represents only sequence and the 'parts list' of an organism. It took more than a thousand years for science to progress from a relatively detailed knowledge of human anatomy to an adequate understanding of physiology – scientists hope that understanding how genomes function will be much faster. This is the goal of the new research field known as *functional genomics*.

The success of genome sequencing was largely due to the development of high throughput DNA sequencing technology, which has radically changed the way biology is carried out and created a systems approach to biology. Similar high throughput technologies are now emerging for functional genomics. Most notable among them is DNA microarray technology, which permits the researcher to make snapshots of gene expression levels of all of the genes in an organism in a single experiment. There are several names used to refer to this

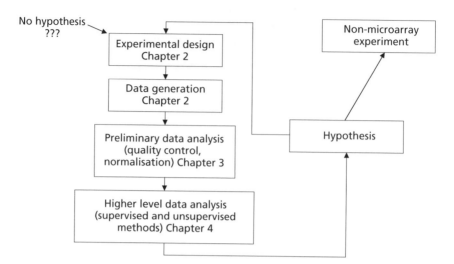

Figure 1.1 An overview of a microarray experiment and data analysis. An experiment is designed, the microarray experiment is carried out, and data are generated. The data must first undergo preliminary processing and quality assessment, and the datasets obtained from different samples have to be normalised before they can be compared directly. Higher level analysis may involve various methods relevant to the biological samples and the information required. The data provide information on RNA expression levels, not on mechanism or causality. Data analysis usually leads to new hypotheses that are tested in follow-up experiments.

technology: DNA microarrays, DNA arrays, DNA chips and gene chips, among others. A distinction is sometimes made between these names but there are no standard definitions for associating particular types of microarray technology with a specific name.

Microarrays are effectively transforming a living cell from a black box into a transparent box. They allow one to identify the genes that are expressed in different cell types, to learn how their expression levels change in different developmental stages or disease states, and to identify the cellular processes in which they participate.

Microarray technology is already producing terabytes of important functional genomics data that can provide clues about how genes and gene products interact and their interaction networks. Unfortunately, transforming these data into knowledge is not a trivial task. The development of methods and tools for the analysis of these huge amounts of complex data is the task of bioinformatics and computational biology. Analysis using multiple techniques is often needed to provide a comprehensive view of the underlying biology. This book provides an overview of data analysis techniques with the aim of facilitating communication between biologists, statisticians and bioinformaticians.

The book consists of three major chapters, which follow this introduction – experimental design, data transformations from raw microarray data to gene expression matrices, and data mining and analysis of gene expression matrices (see Figure 1.1). Chapter 2 covers the principles of experimental design. As with any technology that has the capacity to detect small changes in a highly dynamic system, the underlying experimental design and the manner in which an experiment is conducted is critical for obtaining high quality data. The raw data from microarray experiments are images that have to be transformed and organised into gene expression matrices. These transformations are the subject of Chapter 3. In Chapter 4 we discuss some of the common methods that are used for analysing gene expression data matrices with the goal of obtaining new insights into biology.

Overall the book tries to provide the reader with a general understanding of the nature of microarray data and how it can be analysed. The book does not provide a comprehensive review of the literature, nor is it a reference book describing all existing techniques.

1.1 The central dogma of molecular biology

To understand the essence of gene expression data, we need to consider the 'central dogma' of molecular biology (Figure 1.2). The genetic information of cellular organisms is stored in a long sequence of four different deoxyribonucleotides. These strings of nucleotides are the DNA molecules that compose the genome of an organism. The genome contains segments of DNA that encode genes. Genes are transcribed into messenger RNA (mRNA) and are subsequently translated to form proteins, the main building blocks and functional molecules of a living cell. This process is called gene expression.

DNA is a stable molecule and the same genomic DNA is present, with a few specific exceptions, in all the cells of an organism. Despite this, not all cells are the same. Many of the differences between them are due to the different subsets

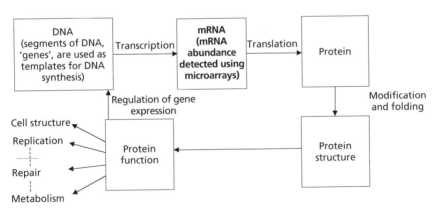

Figure 1.2 The information transfer between DNA, mRNA and protein (the 'central dogma'). Segments of DNA are used as a template to make mRNA, which is used as a template to make protein. The relationship between mRNAs and the genes that encode them can readily be identified, based on the relationship between their sequences. This property is exploited in microarray experiments.

of genes that are expressed in each of the different cell types. We also find different subsets of genes expressed in response to stimuli, so that the pattern of gene expression levels reflects both the cell type and its condition. Microarrays permit the detection of the abundance of various mRNA molecules or *transcripts* in a cell at a given moment. The amount of each mRNA detected in the cell can provide information on the corresponding protein; however, the relationship between the abundance of the mRNA and the corresponding protein is not always straightforward (Gygi *et al.*, 1999).

A DNA molecule consists of two so-called complementary strands, each strand containing the information to determine the other. The RNA molecule transcribed from a gene is complementary to the *coding strand* of the gene. Two complementary single-stranded nucleic acid molecules (i.e. DNA or mRNA) tend to *bind*, or *hybridise*, forming a single, double-stranded molecule. Two single-stranded nucleic acid molecules that are not fully complementary may also hybridise, but the greater the complementarity, the stronger the binding.

1.2 What are microarrays and how do they work?

A microarray is typically a glass or polymer slide, onto which DNA molecules are attached at fixed locations called *spots* or *features* (in the context of microarrays these will be treated as synonyms). There may be tens of thousands of spots on an array, each containing tens of millions of identical DNA molecules (or fragments of identical molecules), of lengths from tens to hundreds of nucleotides. For gene expression studies, each of these molecules should identify a single mRNA molecule, or *transcript*, in a genome. In practice, it is not always possible to identify sequences that monitor the expression of specific transcripts unambiguously, because of the presence of families of similar genes. The features are either printed on the microarrays by a robot or jet, or are synthesised *in situ* by photolithography (similar to the process used in the manufacture of computer chips) or by inkjet printing.

Microarrays may be used to measure gene expression levels in different ways. One of the most popular microarray applications is to compare the gene expression levels in two different samples, e.g. the same cells or cell type under two different conditions (Plate 1.1, facing p. 88). This is based on labelling a representation of the mRNA extracted from each of the samples in two different ways, for instance a green label for the sample from condition 1 and a red one for the sample from condition 2.

The hybridised microarray is excited by a laser and scanned at wavelengths suitable for the detection of the red and green dyes. The amount of fluorescence emitted upon laser excitation corresponds to the amount of nucleic acid bound to each spot. If the nucleic acid from the sample in condition 1 is in abundance, the spot will be green, while if the nucleic acid from the sample in condition 2 is in abundance, it will be red. If both are equal, the spot will be yellow, and if neither are present it will not fluoresce and so appear black. Thus, from the fluorescence intensities and colours for each spot, the relative expression levels of the genes in both samples can be estimated. In this way thousands of data points each providing information about expression of a particular transcript can be obtained from a single experiment.

Other platforms for obtaining gene expression profiles exploit the same principles as those described above, i.e. the specific binding of labelled nucleic acids in solution with a 'library' of immobilised nucleic acids attached to a substrate. One popular variation involves hybridisation of a single-labelled population of nucleic acid to each array. In this instance comparisons are primarily made between the data obtained from different arrays, as opposed to between the labelled populations hybridised to a single array.

Detection of mRNA levels is possible for large numbers of molecules in parallel because of the highly specific preferential binding of complementary single-stranded nucleic acid sequences. This property of nucleic acids was first exploited experimentally as long ago as 1965, and gained widespread acceptance in the form of a technique that became known as the Southern blot (Gillespie and Spiegelman, 1965; Southern, 1975). However, it is only with parallel developments in sequencing of genomes, advances in miniaturisation, and high density synthesis of nucleic acids on solid supports that microarray technology has advanced significantly. Microarrays were first used to study global gene expression in 1997 (DeRisi *et al.*, 1997).

1.3 Gene function and drug discovery

Gene expression studies can be roughly divided in two categories: situations in which samples are used to provide information on genes, and situations in which the genes are used to provide information on the samples. The first approach permits an integrated approach to biology, in which genetic regulation can be examined within the context of circuitry, a sophisticated network in which the interplay of positive and negative signals ultimately directs cellular fate. This approach is already revealing great elegance and efficiency in biological design. The second approach is revolutionising molecular medicine at the level of classification of disease, diagnosis and prognostic prediction, and in a number of industrial and pharmaceutical applications.

Biologists are discovering that genes involved in common processes are often co-expressed. These include genes required for nutrition and stress responses, and genes whose products encode components of metabolic pathways. Similarly, the genes encoding subunits of several multi-subunit complexes such as the ribosome, the proteosome and the nucleosome are also coordinately expressed (Alon *et al.*, 1999; Brown and Botstein, 1999; Causton *et al.*, 2001; Eisen *et al.*, 1998; Hughes *et al.*, 2000; Lashkari *et al.*, 1997). In many cases, this is attributable to coordinate regulation by common factors. 'Waves' of co-expressed temporally regulated genes have also been observed during the development of the rat spinal cord (Wen *et al.*, 1998). Coordinate regulation of genes is extremely efficient, as all the components of multi-subunit complexes or factors required for complex processes are usually required in a defined ratio at the same time, whenever they are needed.

The gene expression profile, or signature, can be thought of as a precise molecular definition of the cell in a specific state (Young, 2000). Accurate, quantitative information on the transcriptional profile of biological samples is therefore of great utility. The expression profile is one of the few relatively accessible ways of describing a phenotype that can be used to characterise a wide variety of samples. Cellular phenotypes can be inferred from gene expression profiles, in part because defects in similar pathways or processes can be detected via their effects on the expression of similar groups of genes, and because agents that perturb these pathways also affect the same gene sets. A large reference collection of profiles against which gene expression data can be compared is therefore useful, but requires careful and accurate data generation, storage and description.

The 'compendium approach', in which large numbers of biological samples are profiled and pattern matching used to predict the function of previously uncharacterised genes and putative drug targets, has been elegantly demonstrated using yeast (Gray *et al.*, 1998; Hughes *et al.*, 2000; Marton *et al.*, 1999). Similarly, databases integrating gene expression data from 60 pharmacologically characterised human cancer cell lines (NCI60, http://dtp.nci.nih.gov/) treated with 70,000 agents independently, or in combinations, have been used to link drug activity with its mode of action, to correlate expression levels of individual transcripts with mechanisms of drug sensitivity and resistance, and to examine the variation in gene expression patterns between individuals. The same dataset was also used to classify cell lines in relation to their tissue of origin and to predict drug chemosensitivity or resistance (Ross *et al.*, 2000; Scherf *et al.*, 2000; Staunton *et al.*, 2001; Weinstein *et al.*, 1997).

Gene expression has proved a highly robust 'reporter' of biological status for a wide range of samples under a variety of conditions, with the result that microarray technology is now utilised extensively within industry. Pharmaceutical companies use microarray technology at numerous stages of drug development, from high throughput screening of small molecules for identifying possible drugs, to drug target identification and assessment of toxicity.

Gene expression data have proven highly informative of disease state, particularly in the area of oncology, where accurate and early diagnosis, followed by appropriate treatment, can prove critical. Studies on clinical samples have shown that gene expression data can be used not only to distinguish between tumour types, define new (histologically indistinct) subtypes, and identify misclassified cell lines, but also to predict prognostic outcomes (Alizadeh *et al.*, 2000; Bittner *et al.*, 2000; Golub *et al.*, 1999; Perou *et al.*, 1999; Shipp *et al.*, 2002). This approach is particularly powerful in offering the promise of 'personalised medicine', in which the specific underlying defect can be identified, the prognosis predicted, and treatment tailored to the genetic makeup of the individual and the specific defect in each patient, thus reducing the likelihood of unwanted side effects.

1.4 Data generation, processing and analysis: an overview

Every high throughput experiment consists of two major parts: (i) *material processing* and *data collection*; and (ii) *information processing*. In a microarray experiment material processing and data collection can be broken down into five steps:

1 array fabrication;
2 preparation of the biological samples to be studied;
3 extraction and labelling of the RNA (or a representation of the RNA) from the samples;
4 hybridisation of the labelled extracts to the array;
5 scanning of the hybridised array.

The scanned image is the starting point for information processing.

Information processing can also be broken down into distinct stages:

6 image quantitation – locating the spots in the image and measuring their fluorescence intensities;
7 data normalisation and integration – constructing the gene expression matrix that describes gene expression values from sets of spot quantitations from different hybridisations;
8 gene expression data analysis and mining, e.g. finding differentially expressed genes or clusters of similarly expressed genes;
9 generation from these analyses of new hypotheses about the underlying biological processes.

The last step, if successful, stimulates new hypotheses that in turn should be tested in follow-up experiments. Note that the material processing steps are preceded by information processing in the *experimental design*. Array design is not a trivial problem, involving, amongst other factors, the selection of an appropriate clone-set or the design of representative nucleic acid sequences that are spotted, or synthesised, on the array. Steps (1)–(5) also require information processing, which can be partly carried out using a laboratory information management system (LIMS). In this book we concentrate mostly on information processing steps (7) and (8). Nevertheless, the basic principles of experimental design are also discussed, since the design can significantly affect the character and the quality of the data that can be used for the subsequent information processing.

The raw data that are produced from microarray experiments are digital images. To obtain information about gene expression levels, these images are analysed, each spot (feature) on the array identified and its intensity measured and compared with values representing the background. Image quantitation is usually carried out using image analysis software. Although image analysis is still considered one of the bottlenecks of microarray technology, significant improvements have been made during the past few years. Image analysis can now be regarded as an area for experts, and comprehensive knowledge of the details of image analysis may not be necessary for using the software. However, it is useful to know some principles of image analysis for better understanding the nature and limitations of microarray data.

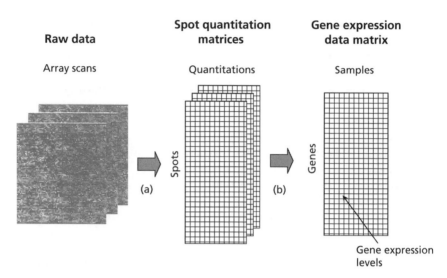

Figure 1.3 Processing of raw data into a gene expression matrix. Data transformation consists of two steps: (a) image quantitation and (b) normalisation and the combining of replicate measurements. In spot quantitation matrices rows typically represent all of the measurements made from individual spots on the array. An experiment typically consists of one or more spot quantitation matrices representing all of the arrays used in the study. In the gene expression matrix, columns represent individual arrays and rows represent the genes and their measurements across all of the arrays. (Reprinted with permission from *Nature Genetics* (Brazma *et al.*), Copyright 2001 Nature Publishing Group.)

An essential feature of all image analysis software is that the digitised microarray images are processed and the data are extracted and combined in a table. This is known as a *spot quantitation matrix* (see Figure 1.3). Each row corresponds to one spot on the array, and each column represents different quantitative characteristics of that spot, such as the mean or median pixel intensity of the spot and local background. Some aspects of image analysis are discussed in Chapter 3.

Generation of the spot quantitation matrices is only an intermediate stage in data processing. The data from multiple hybridisations must be further transformed and organised in a *gene expression matrix*. In this matrix each row represents a gene, or transcript (as opposed to a feature on the array), and each column represents an experimental condition, such as a particular biological sample. Each position in such a matrix characterises the expression level (relative or absolute) of a particular gene, or transcript, under a particular experimental condition. In addition to information on gene expression, we would also ideally like to have data that could be used to characterise the reliability or accuracy of each measurement, e.g. the standard deviation of replicate measurements.

Obtaining the gene expression matrix that combines the information from multiple spot quantitations is again not a trivial task. First, a single gene can be represented by several features on the array, containing the same, or different, sequences. Second, the same experimental condition can be monitored in multiple hybridisations carried out over replicate experiments. All the quantities relating to a gene (either on the same array or on arrays measuring the same conditions in replicated experiments) have to be combined to obtain a single number. Moreover, measurements obtained using different arrays, or even by different locations on the same array, have to be normalised to make them directly comparable. Microarray data normalisation and the treatment of replicate measurements are still a major focus of microarray data analysis research. There are no standard best methods fitting all cases, and new, more sophisticated methods are being developed, so the researcher has to understand these methods in some detail to be able to make appropriate choices. Data normalisation and the treatment of data from replicates are therefore a major focus of this book, and are discussed in Chapter 3. The gene expression matrix may also be annotated to include additional information about the genes or the experimental conditions that is not directly represented in the matrix itself.

After the annotated gene expression data matrix has been generated, we can begin analysing and mining it (step 8). Gene expression levels described in the matrix are typically not measured in standard or objective units. In most cases the numbers are comparisons of gene expression of the given experimental condition to a reference sample, which may or may not be the same throughout the matrix. This complicates data analysis considerably.

The simplest way of analysing gene expression data is to identify genes that are differentially expressed in two given samples. Here the main problem is in finding the confidence threshold at which the difference in gene expression can be considered significant. The only really reliable information here is based on replicate experiments, although some heuristic methods can be used. There are two general types of replicates: biological and technical. Biological replicates use independently derived samples to permit assessment of the variability between individual samples, as well as that inherent in the assay. Technical replicates use repeated measurements of the same samples, either within the same array or in independent hybridisations. Both can help to improve experimental measurements and in the identification of differentially expressed genes, however biological replicates are considered to provide more information.

Data analysis is based on the hypothesis that there are biologically relevant patterns to be discovered in the data. For example, there may be genes whose pattern of expression allows the samples to be classified, or that reflect specific cellular responses. The data mining process typically relies on analysis of the gene expression matrix using unsupervised or supervised methods. The latter involves the use of additional information such as that obtained from partitioning of known samples into healthy and diseased categories. Clustering is an example of an unsupervised method and class prediction is an example of a supervised method used in gene expression data analysis. It should be noted, however, that these methods are only tools for biologically meaningful data analysis. An understanding of these methods is needed to choose those most appropriate for solving a particular problem. Gene expression data analysis and mining is discussed in detail in Chapter 4.

One of the approaches to data analysis downstream of clustering is the reverse engineering of gene regulatory networks (Laub *et al.*, 2000). This is based on the hypothesis that genes that have similar expression profiles under a variety of conditions are likely to be regulated by common mechanisms. Therefore, if genes are clustered based on the similarities in their expression profiles and the promoter sequences are obtained for genes within such clusters, some of these promoter sequences may contain a 'signal', e.g. a specific sequence pattern, relevant to the regulation of these genes. This has been used to discover putative promoter elements in yeast genome (e.g. Vilo *et al.*, 2000). There are no rules for this type of data analysis, and the results can produce only further hypotheses that have to be verified in additional laboratory experiments.

Application of different algorithms, or even different parameters (such as distance measures using the same algorithm), or different data filtering methods can produce different results. Some may be artefacts of the methods used; others may reflect the fact that cells typically carry out multiple processes simultaneously via multiple interacting pathways.

Expression data analysis methods are currently only in their infancy. Even the more obvious approaches, such as cluster analysis and the identification of differentially expressed genes, have been used only crudely. For instance, the appropriateness of gene expression profile similarity measures has not been explored systematically and they are typically used in an *ad hoc* manner. Information characterising the quality or reliability of different data points is frequently absent. In the next generation of microarrays, where each spot is printed or synthesised multiple times, it will be significantly easier to estimate the measurement reliability using the standard deviation between the individual measurements, and this information may be used in subsequent data mining.

As with genome sequencing, systematic gene expression profiling is not an end in itself, but rather a tool for creating infrastructure for further research. There is a long way between having detailed gene expression profiles and real understanding of the underlying cellular processes. Bioinformatic methods and tools will be needed to cope with the huge amounts of data, but they will not bring deep understanding by themselves. On the other hand, traditional 'gene by gene' methods will not be sufficient to understand gene regulatory networks consisting of thousands or tens of thousands of interdependent genes. Hypothesis driven and data mining approaches need to be used hand in hand with high throughput data analysis.

1.5 Data management

Data generated in microarray experiments form a powerful resource if, like genome sequence data, they are carefully recorded and stored in databases, where they can be queried, compared and analysed using different computer software programs. A gene expression database consists of three major parts –

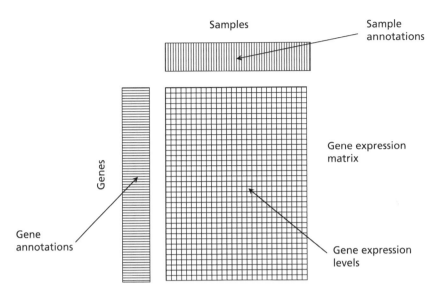

Samples

Sample
annotations

Genes

Gene expression
matrix

Gene
annotations

Gene expression
levels

Figure 1.4 Conceptual view of a gene expression data matrix. There are three parts to a gene expression data matrix: (i) the gene expression data matrix; (ii) gene annotation; and (iii) sample annotation. The gene and sample annotation are important as the data only have meaning within the context of the underlying biology. (Reprinted with permission from *Nature Genetics* (Brazma *et al.*), Copyright 2001 Nature Publishing Group.)

the gene expression data matrix already described, gene annotation and sample annotation (Figure 1.4).

Gene annotation can be provided, to some extent, by links to sequence databases. This is complicated by the many-to-many relationships between genes in the gene expression matrix and features on the array, that make it necessary to have a full and detailed description of each of the features. The lack of consistency in the naming of genes is a serious difficulty, though not one limited to microarray research. A table relating each feature on the array to a list of all synonymous names of the respective gene is a valuable resource.

Microarray technology is still developing rapidly, so it is natural that there are no established standards for microarray experiments or processing of the raw data. There are also no standard ways for measuring gene expression levels. In the absence of such standards the details of how the gene expression data matrix was obtained should be stored in the database, if the data are to be accurately interpreted later. The Microarray Gene Expression Data Society (MGED; http://www.mged.org/) has developed recommendations for the 'Minimum Information About a Microarray Experiment' (MAIME), that attempt to define the set of information sufficient to interpret the experiment, and the results of the experiment, unambiguously, and to enable verification of the data (Brazma *et al.*, 2001).

References

Alizadeh, A. A., Eisen, M. B., Davis, R. E., Ma, C., Lossos, I. S., Rosenwald, A., Boldrick, J. C., Sabet, H., Tran, T., Yu, X., *et al.* (2000). Distinct types of diffuse large B-cell lymphoma identified by gene expression profiling. *Nature* **403**, 503–511.

Alon, U., Barkai, N., Notterman, D. A., Gish, K., Ybarra, S., Mack, D., and Levine, A. J. (1999). Broad patterns of gene expression revealed by clustering analysis of tumor and colon tissues. *Proceedings of the National Academy of Sciences of the United States of America* **96**, 6745–6750.

Bittner, M., Meltzer, P., Chen, Y., Jiang, Y., Seftor, E., Hendrix, M., Radmacher, M., Simon, R., Yakhini, Z., Ben-Dor, A., *et al.* (2000). Molecular classification of cutaneous malignant melanoma by gene expression profiling. *Nature* **406**, 536–540.

Brazma, A., Hingamp, P., Quackenbush, P., Sherlock, G., Spellman, P., Stoeckert, C., Aach, J., Ansorge, W., Ball, C. A., Causton, H. C., *et al.* (2001). Minimum Information About a Microarray Experiment (MIAME) – toward standards for microarray data. *Nature Genetics* **29**, 365–371.

Brown, P. O., and Botstein, D. (1999). Exploring the new world of the genome with DNA microarrays. *Nature Genetics (Supplement)* **21**, 33–37.

Causton, H. C., Ren, B., Koh, S.-S., Harbison, C. T., Kanin, E., Jennings, E. G., Lee, T. I., True, H., Lander, E. S., and Young, R. A. (2001). Remodeling of yeast genome expression in response to environmental changes. *Molecular Biology of the Cell* **12**, 323–337.

DeRisi, J. L., Iyer, V. R., and Brown, P. O. (1997). Exploring the metabolic and genetic control of gene expression on a genomic scale. *Science* **278**, 680–686.

Eisen, M. B., Spellman, P., Botstein, D., and Brown, P. O. (1998). Cluster analysis and display of genome-wide expression patterns. *Proceedings of the National Academy of Sciences of the United States of America* **95**, 14863–14868.

Gillespie, D., and Spiegelman, S. (1965). A quantitative assay for DNA–RNA hybrids with DNA immobilized on a membrane. *Journal of Molecular Biology* **12**, 829–842.

Golub, T. R., Slonim, D., Tamayo, P., Huard, C., Gaasenbeek, M., Mesirov, J. P., Coller, H., Loh, M. L., Downing, J. R., Caligiuri, M. A., *et al.* (1999). Molecular classification of cancer: class discovery and class prediction by gene expression monitoring. *Science* **286**, 531–537.

Gray, N. S., Wodicka, L., Thunnissen, A.-M. W. H., Norman, T. C., Soojin Kwon, S., Espinoza, F. H., Morgan, D. O., Barnes, G., LeClerc, S., Meijer, L., *et al.* (1998). Exploiting chemical libraries, structure, and genomics in the search for kinase inhibitors. *Science* **281**, 533–538.

Gygi, S. P., Rochon, Y., Franza, B. R., and Aebersold, R. (1999). Correlation between protein and mRNA abundance in yeast. *Molecular and Cellular Biology* **19**, 1720–1730.

Hughes, T. R., Marton, M. J., Jones, A. R., Roberts, C. J., Stoughton, R., Armour, C. D., Bennett, H. A., Coffey, E., Dai, H., He, Y. D., *et al.* (2000). Functional discovery via a compendium of expression profiles. *Cell* **102**, 109–126.

Lashkari, D. A., DeRisi, J. L., McCusker, J. H., Namath, A. F., Gentile, C., Hwang, S. Y., Brown, P. O., and Davis, R. W. (1997). Yeast microarrays for genome-wide parallel genetic and gene expression analysis. *Proceedings of the National Academy of Sciences of the United States of America* **94**, 13057–13062.

Laub, M. T., McAdams, H. H., Feldblyum, T., Fraser, C. M., and Shapiro, L. (2000). Global analysis of the genetic network controlling a bacterial cell cycle. *Science* **290**, 2144–2188.

Marton, M. J., DeRisi, J. L., Bennett, H. A., Iyer, V. R., Meyer, M. R., Roberts, C. J., Stoughton, R., Burchard, J., Slade, D., Dai, H., *et al.* (1999). Drug target validation and identification of secondary drug target effects using DNA microarrays. *Nature Medicine* 4, 1293–1301.

Perou, C. M., Jeffrey, S. S., van de Rijn, M., Rees, C. A., Eisen, M. B., Ross, D. T., Pergamenschirov, A., Williams, C. F., Zhu, S. X., Lee, J. C., *et al.* (1999). Distinctive gene expression patterns in human mammary epithelial cells and breast cancers. *Proceedings of the National Academy of Sciences of the United States of America* 96, 9212–9217.

Ross, D. T., Scherf, U., Eisen, M. B., Perou, C. M., Rees, C., Spellman, P., Iyer, V., Jeffrey, S. S., Van de Rijn, M., Waltham, M., *et al.* (2000). Systematic variation in gene expression patterns in human cancer cell lines. *Nature Genetics* 24, 227–244.

Scherf, U., Ross, D. T., Waltham, M., Smith, L. H., Lee, J. K., Tanabe, L., Kohn, K. W., Reinhold, W. C., Myers, T. G., Andrews, D. T., *et al.* (2000). A gene expression database for the molecular pharmacology of cancer. *Nature Genetics* 24, 236–244.

Shipp, M. A., Ross, K. N., Tamayo, P., Weng, A. P., Kutok, J. L., Aguiar, R. C., Gaasenbeek, M., Angelo, M., Reich, M., Pinkus, G. S., *et al.* (2002). Diffuse large B-cell lymphoma outcome prediction by gene-expression profiling and supervised machine learning. *Nature Medicine* 8, 68–74.

Southern, E. M. (1975). Detection of specific sequences among DNA fragments separated by gel electrophoresis. *Journal of Molecular Biology* 98, 503–517.

Staunton, J. E., Slonim, D. K., Coller, H. A., Tamayo, P., Angelo, M. J., Park, J., Scherf, U., Lee, J. K., Reinhold, W. O., Weinstein, J. N., *et al.* (2001). Chemosensitivity prediction by transcriptional profiling. *Proceedings of the National Academy of Sciences of the United States of America* 98, 10787–10792.

Vilo, J., Brazma, A., Jonassen, I., Robinson, A., and Ukkonen, E. (2000). Mining for putative regulatory elements in the yeast genome using gene expression data. Paper presented at *Eighth International Conference on Intelligent Systems for Molecular Biology* (American Association for Artificial Intelligence Press, Menlo Park, CA).

Weinstein, J. N., Myers, T. G., O'Connor, P. M., Friend, S. H., Fornace, A. J., Jr., Kohn, K. W., Fojo, T., Bates, S. E., Rubinstein, L. V., Anderson, N. L., *et al.* (1997). An information-intensive approach to the molecular pharmacology of cancer. *Science* 275, 343–349.

Wen, X., Fuhrman, S., Michaels, G. S., Carr, D. B., Smith, S., Barker, J. L., and Somogyi, R. (1998). Large-scale temporal gene expression mapping of central nervous system development. *Proceedings of the National Academy of Sciences of the United States of America* 95, 334–339.

Young, R. A. (2000). Biomedical discovery with DNA arrays. *Cell* 102, 9–16.

CHAPTER 2

Experimental design

Good ideas are not adopted automatically. They must be driven into practice with courageous patience.

Admiral Hyman Rickover

Microarray experiments usually generate large amounts of data. If these experiments are carefully planned and executed the data can be added to and mined over a long period of time, in combination with data from other laboratories. This chapter describes the steps in experimental design that may be taken if microarray data are to be exploited in this way. An overview of the more technical aspects of microarray experiments and detailed protocols can be found in Jordan (2001). Most of the aspects of experimental design discussed here would be part of normal 'good' experimental practice, although some reflect knowledge gained with the benefit of hindsight, from conducting data analysis.

2.1 Experimental objectives and features of microarray data

The first questions to ask before designing a microarray experiment are, 'What are the objectives of the experiment?' and 'Is a microarray experiment the best way to achieve these objectives?' A single microarray experiment produces data on the expression levels of thousands of genes, but has the disadvantage that data from individual hybridisations may be noisy and single data points may not be reliable, particularly for genes with low abundance transcripts. In addition, the genes of interest are not always readily apparent among the data, and large fold changes in the amount of a transcript are not necessarily indicative of greater biological relevance.

Microarray data provide information only about relative gene expression levels in a set of samples, and obtaining reliable data on absolute expression levels from a microarray experiment is difficult. DNA arrays do not measure protein abundance, nor is the correlation between messenger RNA (mRNA) and protein abundance straightforward (Gygi *et al.*, 1999). The data themselves provide no direct information on mechanism, i.e. which of the many differences in expression levels of transcripts are relevant to the biological phenomenon under study and which reflect inherent variability in the system. The use of time courses and conditional mutations are some of the ways in which more information can be obtained from the system under study and can assist the researcher in identifying the genes of interest. Taken together this means that it is important to consider how the data will be combined with that obtained from other sources and to design the experiment to exploit existing knowledge about the system under study, whenever possible.

Microarray technology permits the detection of small differences in transcript abundance and so the data reflect the experimental conditions with exquisite sensitivity. This means that any reduction in the number of variables that are not central to the experimental question being addressed will increase the utility of the data. Some of the experimental variables that can confound a microarray experiment are not usually even considered to be variables. For example, refeeding a population of cells produces a transcriptional response, as does a temperature shift, which might be used to inactivate a conditional mutant. Biological variation is likely to be the greatest source of variability in the output, and the more that can be done to address this in the experimental design, the more informative the data will be.

Microarray data provide information about the overall amount of mRNA in a sample, which may come from a mixture of different cell types, each having different transcript profiles. Therefore differences in mRNA abundance detected using microarrays reflect not only differences in gene expression but also any differences in the composition of the sample. The amount of mRNA can be considered a reflection of the expression level of a transcript only if the samples are relatively homogeneous and the stability of the transcript does not change between the conditions being compared.

In general, microarray data are best used to describe the expression of large sets of genes, as opposed to the expression of a small set of *a priori* known genes (in which case more 'traditional' techniques such as Northern blotting may perform better). Microarray experiments may help to identify *a priori* unknown sets of genes that characterise particular disease states or that permit distinction between particular subtypes of different disease states. Microarray technology has made exploratory, rather than hypothesis driven, approaches to biology possible; however, these approaches are most effective when based on large datasets or when used on well-characterised biological systems.

2.2 General principles of experimental design

A microarray experiment consists of one or more hybridisations each of which involves at least one *labelled extract* (the population of nucleic acid in solution) and one array. The same extract can be hybridised to multiple arrays, or several extracts (usually with different labels) can be hybridised to the same array. In the terminology used here, the labelled extract is a representation of the *sample* (Figure 2.1). A sample can be described in terms of the origin of the biological material – the *biosource* – and the *treatment*.

In this section we discuss issues related to the general experimental design, such as reducing the number of non-essential variables, determination of the optimal number of hybridisations, choice of a reference extract, and whether to use external controls. Topics related to the choice and preparation of the samples and arrays are discussed in the sections that follow.

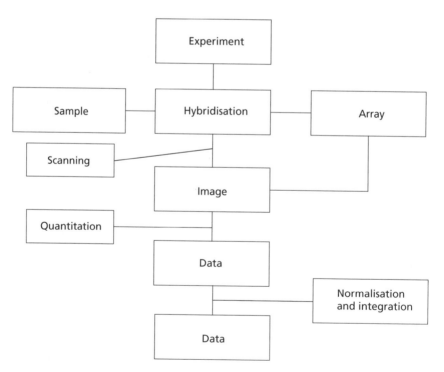

Figure 2.1 Conceptual view of a microarray experiment. A microarray experiment consists of one or more hybridisations, each of which relates at least one labelled extract and one array. The hybridised array is scanned and the image obtained is analysed to obtain gene expression data.

Many of the topics discussed below would be the same for designing any experiment in which the system is very sensitive to small perturbations. This means that adherence to very basic principles, such as those of changing only one variable at a time and characterising the system (including recording as much detail as possible about the experiment – this includes information not normally considered relevant), will pay off when it comes to data analysis.

2.2.1 Reducing the number of variables

A large number of environmental factors – such as temperature, light, time of sampling, humidity, number of animals per cage and the person conducting the experiment – may contribute to differences in transcription. Batch to batch variation in bottles of culture media and brand differences in reagents have the capacity to effect subtle changes in gene expression and can complicate data analysis, especially where the number of samples is small.

The underlying experimental design can often be structured to reduce the number of non-essential variables. For example, if a single culture is split immediately prior to treatment, or cells are harvested at different time intervals from a single culture, between-flask variation will be less than if cells from multiple flasks are processed in parallel.

There are situations where pooling of samples has proved useful for reducing the apparent effects of experimental artefacts. For example, sister transfectants are often very different from each other, more different from each other than primary cells taken from different individuals. This is probably due to gross changes that occur in culture, such as chromosomal rearrangements; in this instance loss of information about individual populations of cells can be an advantage. Another situation in which pooling has proved useful is in the analysis of tissue samples from a single individual, as there is great variability in the distribution of cell types, even in relatively homogeneous tissues such as muscle. Here, pooling of equal amounts of independently derived labelled extracts has been used successfully to reduce the effects of variability in individual samples (Bakay *et al.*, 2002; Chen *et al.*, 2000).

Pooling has the advantage that variation between individual biosources is diluted, although it does not amplify common patterns of gene expression. The more different biosources used, the less each individual contributes to the overall population of labelled extract, such that abnormalities from one atypical sample are less likely to skew the RNA population significantly. If pooling can be avoided, the same experiment carried out on multiple individual samples is likely to be significantly more informative and can provide additional information on the variation between samples and the common elements of the phenomenon under study, and thus the opportunity to assign an additional statistical measure of confidence in the result obtained. As with the issue of replication of samples, the decision on whether to pool, and the optimal number of samples, depends on the inherent variability between samples, the confidence required in the result, and the nature of the downstream data analysis.

Wherever possible, pooling to reduce variability should be carried out late in sample preparation, for example once the labelled extract has been prepared and equal amounts of each extract and equal numbers of extracts used to generate each pool. Pooling is discussed again, in connection with approaches to small amounts of sample, in Section 2.3.2.

The stress response is induced in response to large numbers of stimuli, is fairly general for many types of stress, and involves large numbers of genes (Causton *et al.*, 2001; Gasch *et al.*, 2000). This means that conditions for microarray experiments should be chosen carefully. Less extreme conditions may result in more informative data if the phenomenon of interest is likely to be obscured by stress response genes, or to result in cell death, while very mild conditions may not induce the expected response.

2.2.2 Time courses vs. independent data points

Microarray experiments are frequently used for carrying out pair-wise comparisons between samples at a single time point, e.g. wild type and mutant, treated and untreated cells. The problem with this experimental design is that it is hard to establish which of the differences observed between the two samples is a primary effect, e.g. of the mutation or treatment under study, and which represent a more general accommodation of the cell to life under the experimental conditions, particularly if large numbers of genes are affected.

Time course experiments have proved useful in a number of experimental systems, providing information about the order of events, time scales and trends, as well as confidence that differences in the amount of each transcript detected over the multiple time points reflect changes in the system. Unless the system is well characterised, it is usually hard to know, in advance, at which times cells should be harvested. Approaches to this problem include further characterisation to identify the timing of the phenomenon of interest, for example by monitoring the transcript levels of a few genes of interest over a time course. Another strategy is harvesting cells at a large number of time points, preparing labelled extract (or storing the cells at –80°C) and hybridising and scanning samples from only a subset of the time points, in the first instance. This is a useful strategy when using arrays that are expensive, because much of the variation between labelled extracts reflects biological differences between samples as opposed to differences in preparation of the extract and in the hybridisation. Reasonably comparable data can be obtained using this approach, and it is usually more cost effective than repeating the whole experiment.

2.2.3 Replicates and repeated measurements

Replicate data can be used to provide an estimate of the non-systematic error associated with a measurement. This permits assessment of the significance of results and assignment of a 'confidence score' and can thus increase the precision of estimation. The answer to the question of how many replicates is optimal depends on the purpose of the experiment, the number of variables, how the data will be used, and the confidence required in individual measurements. The high cost of microarray experiments is often used as an argument to limit the number of replicates, but should, on the contrary, be used as an incentive to maximise the amount of information that can be obtained from each experiment. In some cases this may mean carrying out many replicates, in others few, or no explicit replicates at all.

The purpose of generating replicate datasets is generally to assess the range of variability associated with the measurements. There are several potential sources of variability, which fall into at least three major classes: (i) the measurement error (experimental noise); (ii) the natural variability of the measured attribute in the biological system; and (iii) the known or unknown variability in the experimental conditions. Variability in the experimental conditions (except for the experimental factors to be tested) should be reduced as much as possible, as discussed above, so that the known variability may be attributed to the experimental noise and natural variability. The unknown variability in the experimental conditions is difficult to assess and often (unintentionally) is attributed to the natural variability.

Replicates should be chosen to allow separation of the measurement error from the natural variability in the biological system. In a good experimental design the experimental noise should be considerably smaller than the natural variability. In short, the experimental noise should be minimised as much as possible, while the biological variability should be assessed (within the accuracy permitted by the experimental noise), and both should be taken into account in the subsequent data analysis.

There are two different types of replicates: *technical replicates*, which can be used to assess the experimental noise, and *biological replicates*, which are mostly directed towards assessing the natural variability in the system. Technical replicates provide information on the differences between samples from the point at which an individual sample was separated and processed independently. For example, a single-labelled extract of nucleic acid, divided into four tubes and hybridised to four individual arrays in parallel, provides information on the between-array and downstream differences, including variability in hybridisation, washing, scanning, detection, etc., but does not provide information on the stages that came before, such as RNA preparation and labelling.

Technical replicates are particularly important during tuning and testing of experimental protocols. There are many levels at which technical replication is possible, including RNA extraction, labelling, hybridisation, and at the array element (reporter) level (in general, replication is possible at each step in the experimental protocol, Figure 2.2). After the experimental protocols have been optimised and the experimental noise has been assessed, data from technical replicates can be used as quality controls for reproducibility, or to obtain extra information for minimising or identifying the noise in subsequent datasets.

Two levels of technical replication are becoming standard in microarray experiments: dye swapping for experiments in which multiple extracts are hybridised to each array (i.e. all of the extracts are labelled with each of the labels,

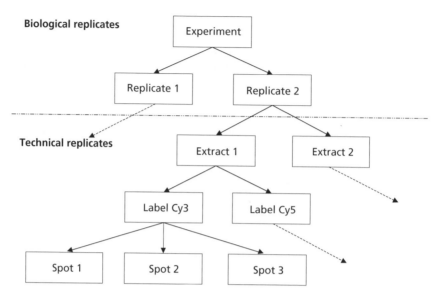

Figure 2.2 Tree representation of replicate experiments. In this example replication is carried out at four levels. The first is at the level of biological replicates. This is followed by two independent mRNA extractions, and reciprocal Cy3 and Cy5 labelling. Finally, on each array each reporter is printed in triplicate. In this way, each data point in the experiment is replicated a total of 24 times. In practice, after the protocols have been tested most of the inherent variability should be biological in origin and some of the replicates may not be needed. Nevertheless, printing reporters in duplicate (or more) and using reciprocal labelling is highly advisable where more than one label is used. Note that technical replicates include the variability that is derived from the biological sample.

e.g. with Cy3 and Cy5, commonly used fluorescent dyes) and replication of reporter sequences on the array. Three or more replicate reporters may simplify image analysis, enabling one to discard the signal from a reporter that has been identified as faulty, and to rely on the data from the remaining reporters instead. Dye swapping is important as different labels may label transcripts with different efficiencies or have different hybridisation kinetics.

Biological replicates are obtained using two or more biological specimens taken from the same biosource and treated in identical ways. The type of replicate needed is determined by what the experimenter wants to obtain from the replicate data. This is often an estimation of the variability in the dataset as a whole, in which case the replicates should reflect only as much of the difference as is expected to be found between samples in the dataset; e.g. if all the cells used for sample preparation are harvested on the same day, cells for a replicate dataset should also be harvested on the same day and processed in parallel with the other samples. Replicate datasets generated using cells harvested on different days are likely to encompass greater variation than that reflected in the dataset under study.

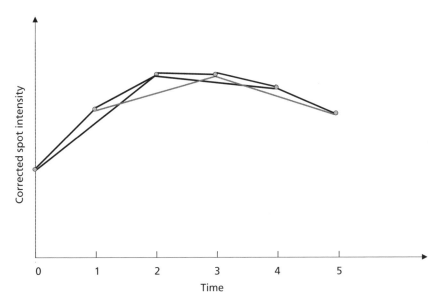

Figure 2.3 Graph showing the changes in gene expression levels plotted against time. Irrespective of whether we use all the time points, or only the even time points, we obtain essentially the same plot. Therefore each of the measurements in this time course provides supporting evidence about the reliability of the neighbouring measurements.

It has been demonstrated for yeast (Hughes *et al.*, 2000) that under (intended) identical conditions the variability in expression levels of different genes may be different for different genes, as some transcripts are present at relatively constant levels, while others have high natural variability in expression level.

The natural variability of gene expression levels should be taken into account if we need to set thresholds for identifying gene differential expression (up- or down-regulation). When comparing two samples, in the absence of any extra information, in general it is impossible to tell from only one measurement whether a given gene is differentially expressed or not, unless various assumptions are made. Therefore replicates are necessary in experiments designed to identify differential gene expression (although information from previous experiments may also be used). On the other hand, some experiments, like time courses, involve a certain amount of implicit replication, even if explicit replicates are not included (Figure 2.3); this is because data from each time point are related to the data from the adjacent time points. In these situations replicates may not be essential, provided that the technology is reliable (or it may be sufficient to have explicit replicates for only some of the time points). However, the possible pitfalls of not carrying out explicit replicates should be considered, and not generating replicates should be a conscious decision. Finally, the way in which the replicate measurements are going to be used should be taken into consideration from the beginning of the experiment.

2.2.4 Reference samples

Microarray data are often used to provide information on the relative 'fold change' in gene expression between two, or more, labelled extracts. The sample with which comparisons are made is usually referred to as the 'reference' sample. The choice of the reference is important, as genes whose expression is activated, or repressed, to the same extent as that in the reference sample may not be identified. For technologies in which a single-labelled extract is hybridised to each array the data used as a reference can be chosen from those within the dataset after the experiment has been carried out, but for array technologies that employ multiple labelled populations of nucleic acid per array it is usually decided as part of the experimental design.

Use of a common reference sample means that data from multiple arrays can be compared with each other. In effect, this is very similar to the situation where a single-labelled population of nucleic acid is hybridised to each array and the data represent absolute relative expression levels. The reference sample must be identical for all the extracts to be compared and should contain all of the genes under study. Examples of common reference samples are genomic DNA or arbitrary pools of RNA. The use of a single reference sample for large numbers of experiments presents some problems; e.g. it can be hard to make enough of a reference sample for comparison of large numbers of arrays over a long time period. 'Spiked-in' controls may also be considered a type of reference sample. This concept was exploited by Dudley *et al.* (2002), who synthesised a reference sample composed of labelled oligonucleotides that hybridised to every feature on the array. The reference sample was used to extend the linear range for signal detection and to obtain information on transcript abundance. Spiked-in controls are discussed in more detail in Section 2.2.5.

Kerr and Churchill (2001) have proposed the use of reference-less *loop designs* for microarray experiments involving hybridisation of multiple labelled extracts to each array. In a typical microarray experiment using a *reference design*, most of the data collected reflects the representation of genes in the reference sample – typically the least biologically interesting sample in the study. Further, since each sample is compared with the reference, all comparisons between the query samples must be inferred, which is less precise than a direct comparison.

In a loop design, samples are systematically compared directly with each other, with an emphasis on comparing the samples that are most closely related, such as adjacent points in a time course experiment. There are a number of advantages to this approach. In such a design, each hybridisation measures expression in a biologically interesting sample, allowing more relevant data to be generated with each assay. In addition, direct comparisons reduce variation in each measurement, allowing expression to be estimated more precisely. However, there are also disadvantages to this approach. Since each sample is compared with multiple others, loop designs can require more RNA than do reference designs, and this can be a problem in experiments analysing rare samples, such as clinical specimens. Further, a single poor quality hybridisation can introduce enough uncertainty and missing data into the analysis that estimating the expression in the other samples can be difficult. Replacing a low quality RNA source can also be difficult, as this would require both a replacement for the bad sample and additional RNA from all samples with which this sample had been hybridised. Finally, loop designs are not easily extensible as adding samples requires a redesign of the experiment and selection of the hybridisation pairs. This can be particularly difficult in prospective studies of clinical samples, where hybridisation assays are done using samples as they become available. Despite these disadvantages, loop designs can give very precise assessments of gene expression levels and should be considered when developing an experimental strategy. The use of loop designs is also discussed in Chapter 3 (Section 3.4.4).

2.2.5 Exogenous ('spiked-in') controls

Exogenous controls are usually RNAs that are added ('spiked in') to a sample at a stage in preparation of the labelled extract. These RNAs are either 'spiked in' to the sample at an early stage in sample preparation and are labelled in the same reaction as the endogenous genes, or are labelled nucleic acids that are added after the endogenous genes have been labelled. The way in which the data obtained from the spiked-in controls is used depends on when these exogenous controls are added. The most common applications are for quality control of a stage in sample preparation (such as generation of copy DNA (cDNA), copy RNA (cRNA) or labelling), or hybridisation) (Baugh *et al.*, 2001), assessing the sensitivity with which transcripts can be detected (Hill *et al.*, 2000), estimation of absolute expression levels, for normalisation (i.e. within- and/or between-extract comparison) (Hartemink *et al.*, 2001; Hill *et al.*, 2001; Holstege *et al.*, 1998; Schuchhardt *et al.*, 2000), to identify specific features on the array (e.g. for the purposes of relating the scanned image to individual features) or for extending the range under which signal intensities can be measured accurately (Dudley *et al.*, 2002).

The controls either represent sequence complementary to that on the array, or result in the synthesis of sequence complementary to that on the array, and are added to the sample in known quantities. For example, bacterial genes from *Bacillus subtilis* are commonly used as external spikes for experiments involving non-bacterial samples, and genes from *Arabidopsis thaliana* for experiments involving non-plant samples. The arrays used must also have features designed to hybridise to the spiked-in controls.

The amount of control added to each sample is known, which can help to relate the spot intensity to the mRNA abundance. This relationship is extrapolated to all the spots on the array and the information can be used for normalisation (to correct for differences within and between labelled extracts), as described in Chapter 3. Spiked-in controls have proved to be of particular utility in situations where the commonly used assumption, that the overall abundance of mRNA from each sample is the same, is not valid. An example is when the overall mRNA population is very different in the extracts under comparison, e.g. when transcription of mRNA is inhibited. In this instance controls for normalisation may be added after total RNA has been prepared. The total amount of RNA in each sample is assessed, usually by spectrophotometry, and a standard amount of the control RNA is added ('spiked in') per milligram of total RNA used for preparation of the labelled extract. The underlying principle is that variation in the population of mRNA will be negligible relative to the total RNA population since about 90% of total RNA is ribosomal RNA (rRNA), which is relatively stable. The assumption that the signal intensities obtained from these spikes should be constant is therefore effectively an assumption that the total RNA (and not the mRNA sample populations, as for total intensity normalisation – see Section 3.3.1) should be the same between extracts.

For experiments using spotted arrays the same mRNA messages from *B. subtilis* or from *A. thaliana* are sometimes added to samples prior to preparation of the labelled extract. The messages are usually used either as positive controls or in a dilution series, for assessing the sensitivity with which messages can be detected. The widespread use of defined controls and probes will greatly enhance the comparability of spotted array data (Schuchhardt *et al.*, 2000).

The controls most commonly used are a cocktail of five *B. subtilis* messenger RNAs. The Dap, Lys, Phe, Thr and Trp mRNAs are generated by *in vitro* transcription from a linearised plasmid template. Since bacterial messages do not have poly A tails and poly A tails are required for synthesis of labelled extract in many protocols, these mRNAs are often prepared by *in vitro* transcription from a plasmid that contains a poly A tract downstream of the bacterial gene and are purified using an oligo-dT column. The requirement that these genes contain a poly A tract means that low amounts of bacterial contamination of the RNA preparation should not contribute to the signal obtained from the *B. subtilis* messenger RNAs. poly A tails are not required for controls added at later stages in sample preparation.

One disadvantage of these approaches is that the addition of the controls has to be planned ahead, i.e. spiked-in controls are added at a point during preparation of the labelled extract and are ideally added to all the samples to be compared, in parallel. The accuracy of the technique is heavily dependent on the ability of the experimenter to add the controls in the appropriate amounts for each sample population. Note that the five *B. subtilis* controls listed above have been known to cross-react with bacterial messenger RNAs and are therefore not suitable for use in experiments where bacteria have been used to infect mammalian cells, as well as any situation where bacterial genomes are being studied. Other genes, such as those from *A. thaliana*, may be suitable in such cases, as long as the transcripts are represented on the arrays and they do not cross-react with those in the endogenous sample.

2.2.6 Dual labelling/dye swapping

Methods for labelling of extract involve either direct incorporation of (usually) fluorescently labelled nucleotides or incorporation of a group that is fluorescently labelled in a second reaction. Although dye incorporation into the sample RNA populations should, in theory, be equivalent, some transcripts incorporate one dye more readily than the other. In situations where extracts labelled with different dyes are compared, each extract should be labelled with each of the fluorescent dyes. This is particularly important where the object of the experiment is to detect differential expression, especially under conditions where few genes are likely to be differentially expressed. Estimates of the proportion of genes that have a bias in Cy3 or Cy5 incorporation lie at around 0.1%, which may seem small, but in an experiment examining the expression of 10,000 genes this works out at 10 genes. Data from dye-swap experiments are therefore essential for establishing which genes are differentially expressed. This is discussed in connection with replicate datasets earlier in this chapter and in Chapters 3 and 4.

One may note that such dye-specific labelling of transcripts is likely to be more common in direct-incorporation labelling reactions than in the secondary coupling protocols that are now commonly used. The advantage of performing a secondary coupling reaction is that the cDNA or cRNA synthesis reactions are identical for both samples. This reaction incorporates modified nucleotides and is followed by a covalent coupling reaction in which the dyes attach to the modified residues. This eliminates any steric or other bias for incorporation of one dye over the other in reverse transcription, and consequently decrease noise in the expression ratios measured at low intensity. Nevertheless, some dye-specific effects are still present, probably caused by a wavelength-dependent differential response of the photomultiplier tubes in the array scanners or slight differences in the quantum efficiencies of the fluorescent dyes, but these can be removed by application of Lowess normalisation and replicate filtering, as discussed in Chapter 3.

2.2.7 Validation of results

As with any experiment, conclusions, especially those relating to small groups of genes (as opposed to genome-wide expression profiles), should be validated. The extent and nature of validation is usually determined by the confidence required in a result. Validation may be carried out by replication, or by using other techniques. Replication has the advantage that information on the sensitivity and statistical significance of the original data is obtained, but does not always assist detection of experimental artefacts. Non-microarray-based methods address the latter.

More common methods for validating results obtained on individual genes are Northern blots, RT-PCR (real time polymerase chain reaction) or TaqMan assays, which permit estimation of transcript levels. However, microarray technology is now fairly robust and so other techniques that directly or indirectly verify the results of microarray analysis may also be considered, e.g. *in situ* hybridisation, which provides information about the location of a transcript, or techniques that permit examination of the corresponding protein, such as Western blotting or 2D gel electrophoresis.

2.3 Choice and preparation of samples

The choice and preparation of the sample should minimise the number of experimental variables which are not central to the biological question under study.

2.3.1 Obtaining the appropriate sample

2.3.1.1 STRAIN BACKGROUND

The organisms to be compared should be as similar as possible, unless comparison of differences between them is one of the objectives of the experiment. This matters less as the number of samples increases, but wherever possible congenic or isogenic strains should be used and these should be as similar as possible, e.g. of the same sex/mating type, age, have the same markers. The use of arrays to probe differences between strains has been described for yeast and mice (Pavlidis and Noble, 2001; Primig *et al.*, 2000). Research on *Drosophila* shows that gene expression in adults is affected most strongly by sex (7%), then by genotype and then by age (1%), at the 0.0001 level of significance (Jin *et al.*, 2001).

Hybridisation properties are used to provide a measure of gene expression, therefore it is sometimes worth considering how similar the strain used for preparation of the labelled extract is to the strain whose sequence is represented on the microarray. The more similar the strains are, the greater the stringency of hybridisation. The number of genes whose expression can be scored with confidence decreases the more the strains differ at the sequence level (this is of particular relevance in situations where the sequences represented on the array are short, e.g. oligonucleotide arrays). However, samples from species such as chimpanzee and rhesus monkeys have successfully been analysed on arrays where the spots represent sequences from human DNA (Bigger *et al.*, 2001; Kayo *et al.*, 2001).

A few strains have been identified that produce unusual results when gene expression is analysed in comparison with other strains. These include the 129SvEv mouse strain, which harbours mutations in the *Gas5* gene that affect mRNA stability (Muller *et al.*, 1998; Sandberg *et al.*, 2000).

2.3.1.2 MUTANTS

The function of a gene can be explored by comparing samples derived from wild type and mutant forms of the gene. This appears to be an extremely straightforward type of experiment; however, a number of factors should nevertheless be considered. First, differences between the wild type and mutant reflect not only the primary but also the secondary consequences of the mutation, so the more deleterious the effect of the mutation, the harder it will be to interpret the data. One way to study the effect of deleterious genes is to use a conditional mutation, whose effect is apparent only under some conditions. When using conditional mutants it is important that both the wild type and mutant cells or animals are studied under the same conditions. Second, the mutation should be as minor and relevant to the question under study as possible. Point mutations that 'knock out' the catalytic function of the associated gene product are preferable to whole gene deletions (Madhani *et al.*, 1998). Third, the effect of the gene may be detectable only if it is not redundant, so the use of mutants associated with identifiable phenotypes is more likely to provide useful data.

2.3.1.3 REAGENTS

The purity of the reagents used for microarray experiments is possibly more important than for lower throughput experiments, because of the sensitivity of the read-out. Reagents should be as pure as possible and should specifically target the phenomenon of interest. For instance, pleiotropic drugs are likely to reveal pleiotropic changes in gene expression, complicating data analysis. The effects of pleiotropic drugs may be great enough that the phenomenon of interest is masked by the (unintended) induction or repression of other genes. One example is the response of mammalian cells to a bacterial sugar, lipopolysaccharide (LPS). This sugar is found ubiquitously in preparations of even highly purified recombinant protein of bacterial origin, unless rigorous steps are taken to remove it. The presence of LPS is used as a signal to induce the immune/stress response in many mammalian systems. Experiments in which bacterial extracts are added to mammalian cells should therefore be prepared under 'LPS-free' conditions, unless this forms part of the biological system under study (Alleva *et al.*, 2000).

2.3.1.4 SAMPLE AND SAMPLE COMPOSITION

The use of cell lines and primary cells in culture allows the experimenter to obtain relatively homogeneous cell populations and to manipulate them under more regulated conditions, thus providing greater potential for carrying out well-controlled experiments. However, transformed cell lines should be characterised as far as possible before use, as transformation may be accompanied by changes in copy number and/or rearrangements of chromosomes, which may affect transcript levels. As with any experiment conducted on cells in culture, the relevance of the experimental results must be assessed in relation to the 'natural' environment of the biological material.

The use of primary cells *ex vivo* permits the study of cells under more natural conditions, but other factors need to be considered here. For example, marked region-specific differences in gene expression have been documented, for example within adult mouse brain (Sandberg *et al.*, 2000). For some experiments this can be addressed by simply excising the tissue to be analysed as cleanly as possible, with few contaminating cells. For more complex tissues, one should try to characterise the cellular composition of the tissue by carrying out cell sorting (e.g. using fluorescence activated cell sorting (FACS) analysis) and either use tissue for which the proportion of cell types is the same across multiple samples or start with possibly small amounts of homogeneous material and, if necessary, amplify the RNA.

The amount of each mRNA detected in a microarray experiment depends on the composition of the sample. This means that if the composition of the sample changes, for example during differentiation, what is perceived to be a change in gene expression might instead reflect differences in the distribution of the cell types in the sample population. The same is true in situations where there may be contaminating cells, even if the composition of the sample itself does not change.

The population of cells under study does not have to be homogeneous, but should be as consistent as possible for the extracts compared. This is particularly important if the data obtained from the hybridisation of labelled extracts are compared using the assumption that the overall intensity of the signals from each extract should be the same.

2.3.2 Small amounts of sample: RNA amplification and pooling

One approach to obtaining information from only the cells of interest is to start with a smaller, more homogeneous, population of cells, e.g. cells obtained by laser capture microdissection, and to amplify the RNA extracted (Hooper *et al.*, 2001; Leethanakul *et al.*, 2000; Luo *et al.*, 1999; Ohyama *et al.*, 2000). This has the attraction of permitting analysis of small numbers of cells and has the potential to provide information on the differences and similarities between individual cells. Although the technique is successfully exploited in some laboratories, RNA amplification is experimentally challenging, and many of the existing methods have not been shown to result in unbiased linear amplification of the starting RNA population. Some methods have been shown to have a reproducible bias in amplification. The bias is closely related to the amount of starting material and the number of cycles of amplification, so these should be kept constant, even when more starting material is available. In situations where amplified RNAs have been analysed, the majority of over-represented sequences have been found to be AT rich, while mRNAs under-represented contain a high proportion of messages that are likely to form extensive secondary structure. In some cases, amplification at a higher temperature may help to reduce this bias (M. Kenzelmann, personal communication; Ernst *et al.*, 2002). A few methods have been shown to work well for use with microarrays (Baugh *et al.*, 2001; van Gelder *et al.*, 1990; Pabon *et al.*, 2001).

In some situations the labelled extract may be prepared starting with tissue or cells derived from several individuals or biosources. This may be the only way that enough of the starting material can be obtained.

2.3.3 Preparation of the labelled extract

Having harvested cells or tissue under the appropriate conditions, the first step is to inhibit all further RNA synthesis and degradation as quickly as possible. A lengthy harvesting process, or one that involves using conditions different from those of the experiment, can induce a transcriptional response; e.g. harvesting cells at room temperature in a refrigerated centrifuge can induce cold shock if harvesting takes too long. Inhibition of cellular processes is usually achieved by snap freezing in liquid nitrogen, or by the addition of hot phenol or a chaotropic agent.

Preparation of the labelled extract should be carried out in parallel for as many samples as can reasonably be processed. The quality of the initial total RNA or cDNA preparation is a major factor in obtaining high quality labelled extract for hybridisation. The nucleic acid should be 'clean', i.e. free from polysaccharide or protein contamination. A common way to assess this is based on the ratio of the absorbances at 260 nm and 280 nm. (Note that spectrophotometers have a finite linear range, and absorbance readings are proportional to the concentration of nucleic acid only for readings within this range. It is worth establishing the linear range empirically for any spectrophotometer in use.) In addition, the nucleic acid preparation should be of a consistent size range and yield for the protocol used. Very low yields are potentially a problem, as the resulting extract may not provide an accurate representation of the diversity of nucleic acid in the original sample. Different RNase inhibitors have been found to alter the population of RNA in the initial sample selectively, and should therefore be used only after extensive characterisation and if the same RNase inhibitor can be added to all the samples. Another way to remove RNases is to carry out two rounds of extraction with the chaotropic agent. Although the yield of purified total RNA will be lower, this is an extremely effective and reproducible way to quickly remove protein from the sample.

Although a large number of different methods for preparation of total RNA have successfully been used, the detailed methods for subsequent steps of the protocol depend on the individual researcher, the array to which the sample will be hybridised, and the type of analysis that will be carried out.

2.3.4 Assessing the quality of the labelled extract

The quality of the labelled extract can be assessed based either on gross characteristics, e.g. the ratio of the absorbances at 260 nm and 280 nm, the size range and the yield, or on characteristics relating to individual genes. In situations where preparation of the labelled extract involves amplification primed from the 3′ end, and the extract is hybridised to arrays containing composite sequences, the quality of the labelled extract may be assessed from the ratio of the 3′ to 5′ signal for individual transcripts (composite sequences are different reporter sequences, or spots, on the array that represent different regions along the length of the same transcript). This is particularly useful when *in vitro* transcription is used in sample preparation. Affymetrix (http://www.affymetrix.com) produce an oligonucleotide array called a GeneChip™, with composite sequences representing the 5′ and 3′ ends of a number of genes. A ratio of less than 3 : 1 of 3′ to 5′ signal is usually considered reasonable for data generated using these GeneChips™ and the manufacturer's methods for preparation of the labelled extract. This is because transcription is rarely completely processive and initiates from the 3′ end. Amplified RNA usually has a stronger 3′ bias. Another way in which the quality of samle preparation may be assessed is via the use of spiked-in controls. If the cocktail 'spiked in' includes transcripts at a range of concentrations these data can be used to assess the sensitivity with which transcripts can be detected; the controls can also be used to determine the extent to which the sample has been amplified.

2.4 Choice and design of arrays

2.4.1 Choice of array platform

The method of microarray manufacture and the nature of the substrate can be used to categorise arrays. First there is distinction between the so-called *cDNA arrays*, where polymerase chain reaction (PCR) products are typically used as sequences on the array features (spots), and *oligonucleotide arrays*, or *oligo-arrays*, where the features (spots) are made up of oligonucleotides. Oligo-arrays may be either spotted or synthesised *in situ*, i.e. the features can be made from presynthesised oligomers that are spotted onto the array substrate or from oligonucleotides that are synthesised directly on the substrate (typically either by photolithography or by inkjet printing). The array substrate may be either non-porous, typically glass or a polymer, or porous, typically a nitrocellulose or nylon filter.

The DNA sequences that represent the features (spots) on the array are referred to as *reporter sequences*. Multiple, different, sequences that provide information on the expression of the same transcript, usually a single gene or exon, are known as *composite sequences* (or *composite elements*). An example is the set of 11 to 20 'perfect match' oligonucleotides on an Affymetrix GeneChip™ that hybridise to different parts of a single transcript.

The choice of array platform is often determined by accessibility and cost; however, it is worth considering future implications of this decision, as data will be more comparable if the experimental design and platform are the same for the duration of a project. While manufacture of arrays in-house can be attractive in terms of cost and customisation, the amount of time and effort required to set up and characterise a robust system is appreciable and is advisable only if large numbers of microarray experiments are planned. The use of commercial arrays may place restrictions on the type and distribution of the genes represented on the array, although manufacturers are increasingly willing to construct custom arrays for an appropriate price. This option is usually most viable where large numbers of experiments are planned using the same array design, as the cost per array is frequently smaller for large orders.

Another way to classify microarrays is in relation to the number of labelled extracts that are hybridised to each array at any one time. Single-extract array methodologies usually involve the use of commercial arrays. These range from the very simple, e.g. the filters available from ResGen Invitrogen Corporation (http://www.resgen.com/) or Clontech (BD Biosciences, http://www.clontech.com/), to the highly engineered Affymetrix GeneChip™. The corrected spot intensity measurements are used as absolute values and comparisons are primarily made between extracts hybridised to different arrays. A major advantage of this experimental design is that the comparisons between datasets do not have to be decided 'up front'.

Multiple extract per array methodologies have typically been used with arrays spotted with PCR products and the correct spot intensity measurements used as a ratio of the values obtained from the samples hybridised to a single array. However, the distinctions between the methodologies for generating microarray data are blurring rapidly, and the use of oligonucleotide-based arrays with multiple extracts is becoming widespread.

2.4.2 Oligonucleotides vs. PCR products

PCR products offer the advantage of amplification, so that limited template DNA and clone banks can be spotted multiple times after amplification from a single 'source' plate. However, reliance on amplification means that there are potential problems with contamination. This has been addressed in a number of ways, from the design of liquid-handling robots with 'non-contamination arms' that permit only single wells of a multi-well plate to be exposed at a time, to elaborate methods for primer design, so that clones in neighbouring wells cannot be amplified with the same primers. The sequence available for hybridisation is typically in the range of 400 bp to 1 kb.

Oligonucleotides, typically in the range of 25 to 80 base pairs, offer the advantage that it is easier to standardise the sequence represented at each spot, suffer less from the effects of contamination between spots, as no amplification is involved, and can usually be purchased in a form ready for spotting. However, the design of oligonucleotide features means that the sequences representing each spot must be known. The sequence available for hybridisation is typically shorter for oligonucleotides than for PCR products, such that arrays can be designed for the detection of different combinations of exons and exon boundaries (that may result from alternative splicing); however, the data obtained should be carefully controlled for the effects of possible non-specific hybridisation. Redesign of the oligonucleotide sequence representing the feature usually addresses this concern, as does the use of multiple reporters (composite sequences) for each transcript. The information obtained from composite sequences is also useful on arrays in which the features are represented by PCR products.

The following guidelines have proved useful in designing oligonucleotide elements (T. Freeman, HGMP Resource Centre, Cambridge, UK, personal communication):
- length of 50–70 nucleotides;
- avoid sequences that may be able to form secondary structure;
- GC content of 45–65%;
- 3′ bias;
- less than 70% homology to other known genes.

Programs that can be used to assist in reporter design include OligoArray for oligonucleotide elements (http://berry.engin.umich.edu/oligoarray/) (Rouillard *et al.*, 2002). There are also programs for assessing which expressed sequence tags (ESTs; see Glossary) represent unique transcripts; two of these are available from the National Center for Biotechnology Information (NCBI) (http://www.ncbi.nlm.nih.gov/UniGene/) and The Institute for Genome Research (TIGR) (http://www.tigr.org/tdb/tgi/hgi/index.html). Both of these programs are based on clustering of EST data (Hegde *et al.*, 2000).

2.4.3 Replicate, guide and control features

Arrays with multiple features (spots) representing each sequence and multiple features representing each transcript (either identical reporters or composite sequences) should be printed using more than one pin, or jet, to print each sequence, so that factors such as local background, between-pin variation, and spot morphology can be assessed. Schuchhardt *et al.* (2000) have found that there is significant variability in the signal obtained from individual spots as a result of differences in pin geometry. Some of these problems may be solved as the use of inkjet printing becomes more widespread.

After the array has been scanned, the scanned image has to be correlated with information about the spot location and the spot identity. One way to do this is to superimpose a grid on the scanned image. The inclusion of spots that produce a strong fluorescent signal at the corners or edges of the array and at a few asymmetric positions may allow the use of these spots as 'landing lights' for the grid and to orient the array.

Transcripts that are expected, and not expected, to be found in the labelled extract should be represented on each of the arrays. The signal from these control features is used to provide information on the quality (e.g. sensitivity of detection, quality of the labelled extract, contamination) and thus the confidence associated with each measurement. Examples of genes that are expected to be in the labelled extract include housekeeping genes and exogenous control genes. Those that are not expected to be in a labelled extract might include ribosomal RNA genes and those that represent sequences from other (dissimilar) genomes.

2.4.4 Cross-hybridisation

If the labelled extract contains families of genes with high sequence identity (70% or more), the labelled nucleic acid represents a transcript which may hybridise to the spot designed to detect the transcript from a different gene. This is known as *cross-hybridisation*. The result is that some transcripts may appear to be more abundant than they really are. There are no good ways for identifying cross-hybridisation in situations where the nucleic acid sequence making up the element is long (e.g. a PCR product).

In general, it is best to avoid possible cross-hybridisation by designing the reporters so that there is less than 70% similarity for any pair of reporters.

Unfortunately this is not always possible. Another strategy for increasing specificity is to design reporters that hybridise to the 3′ untranslated region downstream of the gene, which is often more characteristic of an individual transcript than the coding sequence. Until all the genomes of interest are sequenced, we are unlikely to be able to predict all possible cross-hybridisation events. In these cases one should try to minimise the effects of cross-hybridisation by increasing the stringency of washing.

There are several methods for establishing the extent of cross-hybridisation. The approach described below has been used extensively with data from GeneChips, using an algorithm developed by Affymetrix. Each gene and expressed sequence tag is represented by eleven to twenty 25mer oligonucleotides on the array that match the sequence of the gene of interest. These are called the 'perfect match', or 'PM', spots. The perfect match elements represent different regions distributed along the transcript, with a bias towards the 3′ end of the coding region and the 3′ untranslated region. There is a 'mismatch' spot beside each 'match' spot that differs from the match sequence by one nucleotide at position 13, i.e. in the middle of the 25mer. Hybridisation to the mismatch sequence is indicative of non-specific hybridisation. The paired match and mismatch spots for each gene are together, so that local background effects are similar for both, although the sequences for each transcript are distributed on the array. This means that topical defects in the array and background effects do not compromise the quality of the signal obtained from all the spots representing a transcript.

The signal intensities obtained from each of the 11 to 20 match/mismatch pairs representing each transcript are used in a weighted voting system to determine (i) whether the transcript can be reliably detected and to assign a confidence, or discrimination, score ('P' for present, 'M' for marginal and 'A' for absent); and (ii) the expression level of each gene or expressed sequence tag. The signal intensity value derived using the match and mismatch sequences is known as the 'average difference' or 'signal' obtained from analysis of the match and mismatch sequences, depending on the algorithm used.

A statistical approach for identification of cross-hybridising sequences, contamination and array artefacts based on the standard error between 'perfect match' and 'mismatch' sequences on oligonucleotide-based arrays has also been proposed (Li and Wong, 2001a,b).

2.5 Hybridisation, scanning and quality control

Differences in the amount of labelled extract hybridised to each array are ultimately corrected for by normalisation; however, it is advisable to standardise the amount used, so that the linear range for fluorescence intensity across a set of experiments is comparable and the normalisation factor is small.

The starting point for a good hybridisation is a good quality microarray. While printing protocols have matured significantly, array printing remains a mixture of art and science. Fortunately, it is relatively easy to check the quality of each print run to assure that the DNA deposition was successful and gain an overall qualitative measure of the print quality. There are a number of nucleic acid stains, including Vistra Green (Amersham Biosciences), Syto61 (Molecular Probes) and SYBR Green (Molecular Probes), that can be diluted and used to visualise both single- and double-stranded DNA using the same confocal laser scanners that are used to assess hybridisation. A good rule of thumb is to stain and scan both the first and last slides in any print run. This will allow an assessment of the consistency of spotting across the run and permit identification of drop-out spotting pens or empty wells in the microtitre plates. This information is particularly useful in experiments designed to detect the presence or absence of gene expression for particular genes, as the DNA stain allows 'empty' spots to be excluded from consideration. In most experiments, arrays with a large number of drop-out spots or inconsistent printing should not be used. While the cost of printed slides is not inconsequential, typically the RNA samples are more valuable than the slides themselves, and subsequent analysis is best when one uses the highest quality data from the best possible slides.

In assessing hybridisations there are no hard and fast rules defining what constitutes a 'good' scan, but there are a number of guidelines that can be used to check that the sample and the array are of reasonable quality, the hybridisation has worked, and the fluorescence intensity values are as expected. Many of the features of a 'well-spotted' array can be seen by visual inspection of the array after hybridisation. The spots should be evenly spaced and regular in shape, with little background fluorescence between them. Some of the issues associated with spot morphology, including the impact of the slide surface, humidity, temperature and media used for spotting, are described by Hegde *et al.* (2000). A bright background can be an indicator that unincorporated label has not been efficiently removed from the sample preparation, or that there is dust on the array.

Another indicator of quality is the number of genes and expressed sequence tags that can be 'scored' with some measure of confidence. This needs to be based on experience and will depend on a number of factors, including whether the sample was prepared from the same strain as the elements represented on the array, whether there was loss of mRNA (e.g. due to a transcriptional shutdown) and the relative abundance of genes and expressed sequence tags represented on the array.

The array construction itself can provide additional, more quantitative measures of the quality of the hybridisation. If the array is unbiased in its construction, so that up- or down-regulated genes are not physically arranged to be in one area or another of the array, and so that a large number of randomly chosen genes are included, the observed hybridisation should be unbiased as well. A simple way to test this is to compute the mean and standard deviation of $\log_2(Cy5/Cy3)$ measured for the features in each subgrid and then compare them with those for the entire array; a simple t-test will quickly identify outlying subgrids. This can be used to assess whether spatial biases exist in the array. The same test can be applied to detect biases in pen groups. A similar, simple, qualitative approach is to quickly do a total intensity normalisation of the array data (see Chapter 3) and then create a representation of the array with the most significantly up- and down-regulated genes colour coded to indicate their expression. A quick visual inspection will identify any regional bias on the array.

2.6 Long-term considerations

On one hand microarray experiments are relatively expensive; on the other, if carefully planned and executed they can produce data that can be used repeatedly in different contexts over a long period of time. There are many examples of mining and reanalysing of microarray datasets, often in combination with new datasets to produce significant new results (Alizadeh *et al.*, 2000; Lee *et al.*, 2002; Ross *et al.*, 2000; Scherf *et al.*, 2000; Shipp *et al.*, 2002; Staunton *et al.*, 2001; Waddell and Kishino, 2000).

It may therefore be advisable to keep some long-term considerations in mind when designing an experiment. There are three basic principles that should guide these considerations:

1 experiments and the data generated should be carefully documented;

2 experimental procedures, protocols and reference samples should be standardised as much as possible;

3 investment in the quality of the data may prove useful, even if it does not seem important for achieving the immediate goals.

Microarray experiments carried out over long time periods, possibly in different laboratories, can be compared only if the experimental conditions and protocols have been standardised.

2.6.1 Record keeping

It is not easy to define the minimum information that needs to be reported about a microarray experiment that would allow it to be interpreted unambiguously. A recommendation towards this goal has been made by the Microarray Gene Expression Data Society (MGED) in a document known as the 'Minimum Information About a Microarray Experiment' (MIAME) (Brazma *et al.*, 2001). In addition to the raw and processed data, this information includes the description of the experiment as a whole (e.g. experimental factors), the annotation of each sample used in the experiment (e.g. species, cell type and source), the reference or a description of the array platform used in the experiment, and the experimental protocols, including those for data transformations.

Recording these data by hand in a lab-book can be a time consuming task, and software packages are being developed by several academic and commercial groups that allow these data to be recorded more easily. Note that much of the information generated by a laboratory information management system

(LIMS) is in fact recorded automatically. Data recorded by a LIMS need to be stored in a well-structured database.

It is important to keep as much raw data from experiments as possible, including images and information on image quantitations. For instance, the absolute corrected spot intensity values for each channel should be recorded in addition to the ratios, as this can provide valuable additional information regarding the expression of the particular gene and the reliability of the measurement (unfortunately often only the ratios of the corrected spot intensities are reported).

2.6.2 Standardisation

The major reason for standardisation of microarray experiments is to make data from different hybridisations performed over long time periods comparable. Unfortunately there are currently few standards for microarray experiments. As a first step it is important to try to use a consistent reference sample. Some of the larger laboratories have standardised their experimental protocols. The MGED website (http://www.mged.org/) provides a resource for reporting microarray protocols, and MGED is working on a standardised way to describe such protocols. ArrayExpress (Brazma *et al.*, 2002), a public repository of microarray data based at the European Bioinformatics Institute (EBI), provides a resource for reporting protocols and for viewing all the protocols that have been used in experiments submitted to ArrayExpress (http://www.ebi.ac.uk/microarray/ArrayExpress/arrayexpress.html).

Even within individual laboratories, a certain level of standardisation will prove advantageous, ensuring that data generated by different researchers, or even the same researcher, over a period of years can be compared at least to some degree. It is worth thinking about experimental design in this context, even at the stage of designing your first microarray experiment. For example, it might be worth choosing to hybridise to an array that contains more than just your genes of interest so that at a later date other researchers can use the same arrays for their experiments. The data generated may ultimately prove to be a useful source of information for a large number of researchers and bioinformaticians. It is useful to evaluate all the considerations described in this chapter in this context, such as choice of genetic background, source of reagents and consumables, and time points at which cells are harvested.

There are several public repositories for gene expression data, which, in time, are likely to serve a role for gene expression data similar to that of DDBJ/EMBL/GenBank for sequence data: ArrayExpress at the EBI, Gene Expression Omnibus (GEO; http://www.ncbi.nlm.nih.gov/geo/) at the NCBI in the USA, and the Center for Information Biology Experimentation Database (CIBEX) in Japan. In addition, a common data exchange format MAGE-ML (http://www.mged.org/Workgroups/MAGE/mage.html) has been developed as part of a collaborative project between MGED and several major software companies.

References

Alizadeh, A. A., Eisen, M. B., Davis, R. E., Ma, C., Lossos, I. S., Rosenwald, A., Boldrick, J. C., Sabet, H., Tran, T., Yu, X., *et al.* (2000). Distinct types of diffuse large B-cell lymphoma identified by gene expression profiling. *Nature* **403**, 503–511.

Alleva, D. G., Pavlovich, R. P., Grant, C., Kaser, S. B., and Beller, D. I. (2000). Aberrant macrophage cytokine production is a conserved feature among autoimmune-prone mouse strains: elevated interleukin (IL)-12 and an imbalance in tumor necrosis factor-alpha and IL-10 define a unique cytokine profile in macrophages from young nonobese diabetic mice. *Diabetes* **49**, 1106–1115.

Bakay, M., Chen, Y.-W., Borup, R., Zhao, P., Nagaraju, K., and Hoffman, E. P. (2002). Sources of variability and effect of experimental approach on expression profiling data interpretation. *BioMed Central Bioinformatics* **3**, 4.

Baugh, L. R., Hill, A. A., Brown, E. L., and Hunter, C. P. (2001). Quantitative analysis of mRNA amplification by *in vitro* transcription. *Nucleic Acids Research* **29**, E29.

Bigger, C. B., Brasky, K. M., and Lanford, R. E. (2001). DNA microarray analysis of chimpanzee liver during acute resolving hepatitis C virus infection. *Journal of Virology* **75**, 7059–7066.

Brazma, A., Hingamp, P., Quackenbush, P., Sherlock, G., Spellman, P., Stoeckert, C., Aach, J., Ansorge, W., Ball, C. A., Causton, H. C., *et al.* (2001). Minimum Information About a Microarray Experiment (MIAME) – toward standards for microarray data. *Nature Genetics* **29**, 365–371.

Brazma, A., Sarkans, U., Robinson, A., Vilo, J., Vigron, M., Hoheisel, J., and Fellenberg, K. (2002). Microarray data representation, annotation and storage. *Advances in Biochemical Engineering Biotechnology, Chip Technology* **77**, 113–139.

Causton, H. C., Ren, B., Koh, S.-S., Harbison, C. T., Kanin, E., Jennings, E. G., Lee, T. I., True, H., Lander, E. S., and Young, R. A. (2001). Remodeling of yeast genome expression in response to environmental changes. *Molecular Biology of the Cell* **12**, 323–337.

Chen, Y.-W., Zhao, P., Borup, R., and Hoffman, E. P. (2000). Expression profiling in the muscular dystrophies: identification of novel aspects of molecular pathophysiology. *Journal of Cell Biology* **151**, 1321–1336.

Dudley, A. M., Aach, J., Steffen, M. A., and Church, G. M. (2002). Measuring absolute expression with microarrays with a calibrated reference sample and an extended signal intensity range. *Proceedings of the National Academy of Sciences of the United States of America* **99**, 7554–7559.

Ernst, T., Hergenhahn, M., Kenzelmann, M., Cohen, C. D., Bonrouhi, M., Weninger, A., Klaren, R., Grone, E. F., Wiesel, M., Gudemann, C., *et al.* (2002). Decrease and gain of gene expression are equally discriminatory markers for prostate carcinoma: a gene expression analysis on total and microdissected prostate tissue. *American Journal of Pathology* **160**, 2169–2180.

Gasch, A. P., Spellman, P. T., Kao, C. M., Carmel-Harel, O., Eisen, M. B., Stortz, G., Botstein, D., and Brown, P. O. (2000). Genomic expression programs in the response of yeast cells to environmental changes. *Molecular Biology of the Cell* **11**, 4241–4257.

van Gelder, R. N., von Zastrow, M. E., Yool, A., Dement, W. C., Barchas, J. D., and Ederwine, J. H. (1990). Amplified RNA synthesized from limited quantities of heterogeneous cDNA. *Proceedings of the National Academy of Sciences of the United States of America* **87**, 1663–1667.

Gygi, S. P., Rochon, Y., Franza, B. R., and Aebersold, R. (1999). Correlation between protein and mRNA abundance in yeast. *Molecular and Cellular Biology* **19**, 1720–1730.

Hartemink, A. J., Gifford, D. K., Jaakola, T. S., and Young, R. A. (2001). Maximum likelihood estimation of optimal scaling factors for expression array normalization. Paper presented at *SPIE BiOS 2001* (Society for Optical Engineering, San Jose, CA).

Hegde, P., Qi, R., Abernathy, K., Gay, C., Dharap, S., Gaspard, R., Hughes, J. E., Snesrud, E., Lee, N., and Quackenbush, J. (2000). A concise guide to cDNA micro-array analysis. *BioTechniques* **29**, 548–562.

Hill, A. A., Brown, E. L., Whitley, M. Z., Tucker-Kellogg, G., Hunter, C. P., and Slonim, D. K. (2001). Evaluation of normalization procedures for oligonucleo-tide array data based on spiked cRNA controls. *Genome Biology* **2**, 0055.0051–0055.0013.

Hill, A. A., Hunter, C. P., Tsung, B. T., Tucker-Kellogg, G., and Brown, E. L. (2000). Genomic analysis of gene expression in *C. elegans*. *Science* **290**, 809–812.

Holstege, F. C., Jennings, E. G., Wyrick, J. J., Lee, T. I., Hengartner, C. J., Green, M. R., Golub, T. R., Lander, E. S., and Young, R. A. (1998). Dissecting the regula-tory circuitry of a eukaryotic genome. *Cell* **95**, 717–728.

Hooper, L. V., Wong, M. H., Thelin, A., Hansson, L., Falk, P. G., and Gordon, J. I. (2001). Molecular analysis of commensal host–microbial relationships in the intes-tine. *Science* **291**, 881–884.

Hughes, T. R., Marton, M. J., Jones, A. R., Roberts, C. J., Stoughton, R., Armour, C. D., Bennett, H. A., Coffey, E., Dai, H., He, Y. D., *et al.* (2000). Functional dis-covery via a compendium of expression profiles. *Cell* **102**, 109–126.

Jin, W., Riley, R. M., Wolfinger, R. D., White, K. P., Passador-Gurgel, G., and Gibson, G. (2001). The contributions of sex, genotype and age to transcriptional variance in *Drosophila melanogaster*. *Nature Genetics* **29**, 389–395.

Jordan, B., ed. (2001). *DNA Microarrays: Gene Expression Applications* (Springer-Verlag, Berlin).

Kayo, T., Allison, D. B., Weindruch, R., and Prolla, T. A. (2001). Influences of aging and caloric restriction on the transcriptional profile of a skeletal muscle from rhesus monkeys. *Proceedings of the National Academy of Sciences of the United States of America* **98**, 5093–5098.

Kerr, M. K., and Churchill, G. A. (2001). Bootstrapping cluster analysis: assessing the reliability of conclusions from microarray experiments. *Proceedings of the National Academy of Sciences of the United States of America* **98**, 8961–8965.

Lee, P. D., Sladek, R., Greenwood, C. M. T., and Hudson, T. J. (2002). Control genes and variability: absence of ubiquitous reference transcripts in diverse mammalian ex-pression studies. *Genome Research* **12**, 292–297.

Leethanakul, C., Patel, V., Gillespie, J., Pallente, M., Ensley, J. F., Koonongkaew, S., Liotta, L. A., Emmert-Buck, M., and Gutkind, J. S. (2000). Distinct pattern of ex-pression of differentiation and growth-related genes in squamous cell carcinomas of the head and neck revealed by the use of laser capture microdissection and cDNA arrays. *Oncogene* **19**, 3220–3224.

Li, C., and Wong, W. H. (2001a). Model-based analysis of oligonucleotide arrays: Expression index computation and outlier detection. *Proceedings of the National Academy of Sciences of the United States of America* **98**, 31–36.

Li, C., and Wong, W.H. (2001b). Model-based analysis of oligonucleotide arrays: model validation, design issues and standard error application. *Genome Biology* **2**, 0032.I–0032.II.

Luo, L., Salunga, R. C., Guo, H., Bittner, A., Joy, K. C., Galindo, J. E., Xiao, H.,

Rogers, K., Wan, J. S., Jackson, M. R., and Erlander, M. G. (1999). Gene expression profiles of laser-captured adjacent neuronal subtypes. *Nature Medicine* **5**, 117–120.

Madhani, H., Styles, C. A., and Fink, G. R. (1998). MAP kinases with distinct inhibitory functions impart signaling specificity during yeast differentiation. *Cell* **91**, 673–684.

Muller, A. J., Chatterjee, S., Teresky, A., and Levine, A. J. (1998). The gas5 gene is disrupted by a frameshift mutation within its longest open reading frame in several inbred mouse strains and maps to murine chromosome 1. *Mammalian Genome* **9**, 773–774.

Ohyama, H., Zhang, X., Kohno, Y., Alevizos, I., Posner, M., Wong, D. T., and Todd, R. (2000). Laser capture microdissection-generated target sample for high-density oligonucleotide array hybridization. *BioTechniques* **29**, 530–536.

Pabon, C., Modrusan, Z., Ruvolo, M. V., Coleman, I. M., Daniel, S., Yue, H., and Arnold, L. J., Jr. (2001). Optimized T7 amplification system for microarray analysis. *BioTechniques* **31**, 874–879.

Pavlidis, P., and Noble, W. S. (2001). Analysis of strain and regional variation in gene expression in mouse brain. *Genome Biology* **2**, 0042.0041–0042.0015.

Primig, M., Williams, R. M., Winzeler, E. A., Tevzadze, G. G., Conway, A. R., Hwang, S. Y., Davis, R. W., and Esposito, R. A. (2000). The core meiotic transcriptome in budding yeasts. *Nature Genetics* **26**, 415–423.

Ross, D. T., Scherf, U., Eisen, M. B., Perou, C. M., Rees, C., Spellman, P., Iyer, V., Jeffrey, S. S., Van de Rijn, M., Waltham, M., *et al.* (2000). Systematic variation in gene expression patterns in human cancer cell lines. *Nature Genetics* **24**, 227–244.

Rouillard, J.-M., Herbert, C. J., and Zuker, M. (2002). OligoArray: Genome-scale oligonucleotide design for microarrays. *Bioinformatics* **18**, 486–487.

Sandberg, R., Yasuda, R., Pankratz, D. G., Carter, T. A., Del Rio, J. A., Wodlicka, L., Mayford, M., Lockhart, D. J., and Barlow, C. (2000). Regional and strain-specific gene expression mapping in the adult mouse brain. *Proceedings of the National Academy of Sciences of the United States of America* **97**, 11038–11043.

Scherf, U., Ross, D. T., Waltham, M., Smith, L. H., Lee, J. K., Tanabe, L., Kohn, K. W., Reinhold, W. C., Myers, T. G., Andrews, D. T., *et al.* (2000). A gene expression database for the molecular pharmacology of cancer. *Nature Genetics* **24**, 236–244.

Schuchhardt, J., Beule, D., Malik, A., Wolski, E., Eickhoff, H., Lehrach, H., and Herzel, H. (2000). Normalization strategies for cDNA microarrays. *Nucleic Acids Research* **28**, E47–e47.

Shipp, M. A., Ross, K. N., Tamayo, P., Weng, A. P., Kutok, J. L., Aguiar, R. C., Gaasenbeek, M., Angelo, M., Reich, M., Pinkus, G. S., *et al.* (2002). Diffuse large B-cell lymphoma outcome prediction by gene-expression profiling and supervised machine learning. *Nature Medicine* **8**, 68–74.

Staunton, J. E., Slonim, D. K., Coller, H. A., Tamayo, P., Angelo, M. J., Park, J., Scherf, U., Lee, J. K., Reinhold, W. O., Weinstein, J. N., *et al.* (2001). Chemosensitivity prediction by transcriptional profiling. *Proceedings of the National Academy of Sciences of the United States of America* **98**, 10787–10792.

Waddell, P. J., and Kishino, H. (2000). Cluster inference methods and graphical models evaluated on NCI60 microarray gene expression data. *Genome Informatics* **11**, 129–140.

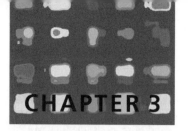

CHAPTER 3

Image processing, normalisation and data transformation

In theory there is no difference between theory and practice. In practice there is.

Andrew S. Tannenbaum

3.1 Introduction

DNA microarrays permit the collection of data on patterns of gene expression by surveying thousands of genes in a single experiment. Each microarray typically contains features representing thousands or tens of thousands of genes that are surveyed in an assay in which one or more samples representing the messenger RNA (mRNA) expressed in the tissues, developmental stages, or treatments of interest are labelled with a distinguishable marker and allowed to hybridise to features on the array. The underlying assumption is that the relative level of detectable hybridisation for each of the labelled extracts for each feature represents the relative population of the corresponding mRNA species in the experimental samples being surveyed. In a typical experiment, one uses these results to identify differentially expressed genes, to find patterns of coordinated gene expression, or to uncover genes whose expression patterns can be used for sample classification.

In glass slide DNA microarray experiments, RNA from the cells and tissues of interest are generally used to generate first-strand DNA (cDNA) complementary labelled with spectrally distinguishable fluorescent dyes such as Cy3 and Cy5. Single-'colour' microarray assays can also be performed and include a single fluorescently labelled sample hybridised to a glass microarray, a radioactively or chemiluminescently labelled sample hybridised to a nylon filter array, or hybridisation followed by labelling with a fluorescent tag, such as that used with Affymetrix GeneChips™. Regardless of approach, the arrays are scanned following hybridisation and independent grey-scale images are generated for query and control samples, typically as 16-bit TIFF (Tagged Image File Format) images.

These images must then be analysed to identify the arrayed spots and to measure the relative fluorescence intensities for each feature. Having identified each spot on the array and having measured its intensity, one must adjust, or 'normalise', the measured hybridisation intensities so that they can be effectively compared within, and between, each array. The data can then be further filtered in order to facilitate effective comparisons.

3.2 Preliminary processing of the data

3.2.1 Image analysis

The first step in the analysis of microarray data is image processing. Most commercially available microarray scanner manufacturers provide software that handles this aspect of the analysis, and there are a number of additional image processing packages available. Nevertheless, it is important to understand how data are extracted from the images as they represent the primary data collected from each experiment and everything else is derived from those images and their initial analysis.

Image processing involves three stages. In the first stage, the spots representing the arrayed features must be identified and distinguished from spurious signals that can arise as a result of precipitated dye or other hybridisation artefacts, contaminants such as dust on the surface of the slide, and other sources of nonspecific background. The problem of finding a distributed collection of features is a difficult one, but for microarrays this is greatly simplified as the systems used to create the arrays generally produce a regular arrangement of features on the surface of the slide. In mechanically spotted arrays, these are typically arranged in *subarrays* or *pen groups* representing each of the pen tips used to deposit the features. Most arrays have their elements arranged in a rectangular grid, although some use an 'orange packing' arrangement to increase spot density.

Microarray image processing software packages generally require the user to specify approximately where each subgrid lies and to provide additional parameters relevant to the spotted array, such as the number of rows and columns in each subgrid, the number and arrangement of subgrids, and the approximate spot size. This information is then used to roughly place the grids over the arrayed spots, adjust them dynamically to best represent the features (often using a 'centre of mass' calculation for each spot), and to determine both the approximate spot area to be surveyed and a local region that can be used to estimate background.

Grid placement is extremely important, as the grid coordinates are used to identify the individual arrayed features and to assign identities to them. Shifting or misaligning the grid can cause particular expression levels to be assigned to the wrong genes. Consequently, many groups use so-called 'landing lights' on their arrays. These are specific features spotted on the array at predetermined locations, typically at the very start and end of each subgrid or around the perimeter of the spotted features being queried in the analysis, that are selected to give a strong fluorescence signal regardless of the quality of the hybridisation. These then provide hybridisation signals that should allow unambiguous placement of the array grid. Landing light features are generally designed to bind to exogenous spiked-in controls (see Section 2.2.5) hybridised to the arrays, polymerase chain reaction (PCR) products incorporating fluorescently labelled nucleotides, or the fluorescent nucleotides themselves.

Once the subgrids have been placed, areas within each individual cell in the grid must be selected to determine the spot signal and an estimate of background hybridisation. There are two widely used approaches for determining the spot area. The easiest is to use a fixed region (for spotted arrays, typically a circle of predetermined area) that is centred on the centre of mass of the spot. This has the advantage that it is computationally simple and provides a reasonable estimate of the area occupied by the spot. A disadvantage is that this method can lead to misestimates of both the spot signal and the local background signal if the spot size is incorrectly estimated or if there is a great deal of variability in the spotted features.

An alternative approach is to attempt to precisely identify the spot boundaries and to include only those pixels within the boundary. This has the disadvantage of being computationally difficult, causing the image processing time to increase and, depending on the particular algorithm used, possibly leading to misidentification of the spot and surrounding background areas. The advantage of this approach is that it allows us to more precisely identify those pixels representing real hybridisation, and consequently it can provide a better estimate of the actual intensity associated with each feature on the array.

3.2.2 Measuring and reporting expression

The starting point for all of our analyses is an estimation of the expression for each gene. One must remember during this process that what we are really doing is trying to infer expression based on measured hybridisation intensities, and that in reality, hybridisation is assaying relative RNA representation, from which we infer expression (Figure 3.1). Consequently, the manner in which hybridisation is measured and reported can have a significant effect on the conclusions that we draw from the experiment.

Once the features have been identified, microarray image analysis software measures the intensities in each channel for each pixel that comprises the image of each feature, and typically reports a variety of summary statistics. These often include the total intensity for each feature and the mean, median and mode of the pixel intensity distribution, as well as an estimate of these for the local background, and other statistics such as the standard deviation of both signal and

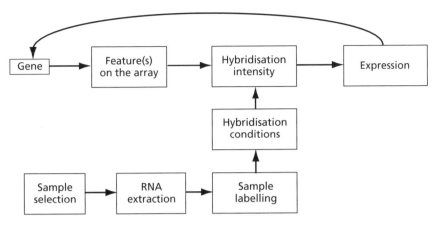

Figure 3.1 Diagram outlining measurement of gene expression. Microarray assays provide an estimate of gene expression. Each gene is represented by one or more features on the array for which fluorescence intensities are measured following hybridisation. From this, we attempt to deduce the expression level of the gene. This requires that we track a range of diverse information, including the samples selected, the conditions under which they are collected, the laboratory protocols and conditions associated with RNA extraction and labelling, and the hybridisation conditions. Ultimately, what we are measuring is RNA representation, not expression, and each step in the process can affect the final result.

background. The primary goal of most microarray assays is to identify differences in expression for each arrayed gene. Typically, we calculate the ratio of signals between independent measurements for each gene, which for two-colour arrays is just the ratio of the measured signal intensity for each channel. There are, however, a number of approaches to measuring the fluorescence intensities for the arrayed features.

Most microarray approaches use either the background-subtracted median or total intensities as the statistic representing each feature. The median, which is the value of the intensity with exactly half of the pixel measurements at greater intensity and half at lower intensity, has the advantage that it is relatively insensitive to outlying, high intensity pixels in the spot image which can be caused by dust, unincorporated label, or other artefacts that cause a small number of saturated pixels within the spot area. However, the median is sensitive to misidentification of the spot area, and an overestimation of the region of the spot that includes too many of the background pixels can skew the median and any quantities calculated from it. This is because the area of the spot, and consequently the number of pixels assessed in calculating the median, grows quickly as the radius of the spot boundary increases beyond the actual area of the feature. The total intensity, on the other hand, is relatively insensitive to misidentification of the spot boundary since adding background pixels with intensities near or equal to 0 after background subtraction has little effect on either the final sum or the final ratio. The total intensity does have the disadvantage that it can be skewed by a few anomalous highly fluorescent pixels that greatly inflate the sum in one or both channels.

3.2.2.1 *SATURATED PIXELS*

Microarray scanners typically use a 16-bit analogue-to-digital converter (ADC) that converts the signal from the photomultiplier tubes (PMTs) that measure fluorescence to a digital value for each pixel in the resulting image. This means that relative intensity levels are reported in a range from 0 to 65,535 ($2^{16} - 1$). The measured values depend on a variety of factors, including the hybridisation itself, the voltage set for the PMT, and the laser power used. In scanning hybridised arrays, one typically tries to strike a balance between detecting as many spots as possible and avoiding saturation of any of the spots.

Regardless, one will typically find a number of pixels within some of the spots that are saturated in one or both channels, and these must be considered and dealt with effectively. One approach is to use the median spot intensity which, as discussed previously, has the advantage of being relatively insensitive to a small number of saturated pixels. However, the median can also be skewed if there are too many saturated pixels in a particular spot. Obviously, pixel saturation will also have a deleterious effect on the integrated intensity. If pixels in a single channel are saturated, this can result in an underestimate of the expression in the sample to which it corresponds. If pixels are saturated in the images representing both labelled extracts, then the result is an unpredictable distortion of the relative expression; examples are shown in Figure 3.2. Consequently, it is generally good practice to eliminate saturated pixels from calculations of spot intensities, and most microarray image processing software allows for this. If there are too many saturated pixels in one or both channels, the spot in that particular image should be flagged as uninformative and eliminated from further consideration.

3.2.2.2 *THE APPROPRIATE NUMBER OF PIXELS*

One other consideration in image processing is the sampling size used for measuring the spots – essentially how many pixels are included in the spot image. For most scanners the default pixel size is 10 μm, so that a circular spot 200 μm in diameter contains about 300 pixels (remember that the area of a circle is $A = \pi r^2$, and the radius of a 200 μm spot is 100 μm, or 10 pixels). This provides a fairly large number of measurements for each spot and allows some confidence that a small number of bad pixels will not unduly influence the result. However, the number of pixels sampled drops rapidly as the diameter of the spot decreases. At 160 μm there are only about 200 pixels in the spot, at 100 μm the number is down to 78, and at 80 μm there are only 50; if any of these are eliminated because they are saturated, or for some other reason, the number of measurements of each spot is further reduced. Recall that the standard error for any measurement grows roughly as $1/n$, where n is the number of measurements, or in our case the number of pixels. This means that as the number of pixels decreases, our uncertainty in any measurement we make grows.

Most scanners now allow use of 5 μm pixels, which provide four times as

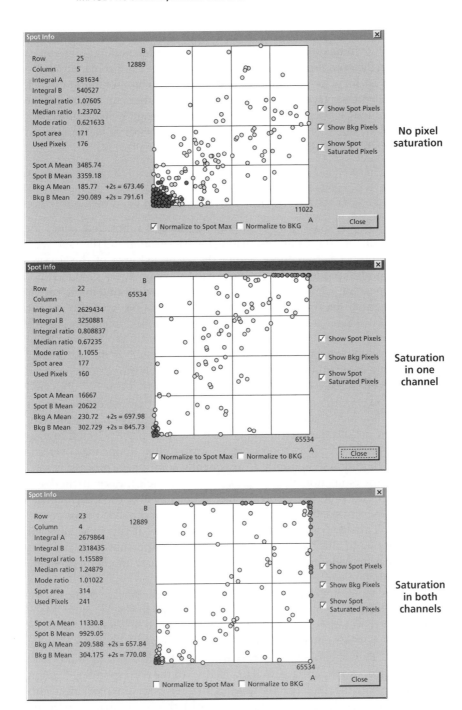

Figure 3.2 Examples of saturation. If pixels are saturated in the images representing one or both labelled extracts, then the result is an unpredictable distortion of the measured relative expression.

many measurements as do 10 μm pixels for a spot of the same size. While it may be useful to use these smaller pixels to increase our confidence in the intensities we measure, their use also increases the image file size by a factor of four relative to that for a 10 μm pixel measurement. Consequently, a balance must be struck between the necessary accuracy of the microarray measurements one would like to make and the practical issues of file storage and image analysis. Generally, 10 μm pixel measurements provide a large enough number of measurements, but for spots smaller than about 80 μm in diameter, smaller pixel sizes should be considered.

3.2.2.3 ESTIMATING BACKGROUND

Most image processing software reports background-subtracted values for each feature on the array by first estimating the background and then subtracting it pixel by pixel from those within the identified area of the feature. However, this first requires an estimation of the appropriate background fluorescence. There are a variety of sources of potential background in each array, including natural fluorescence of the glass or its coating, and non-specific hybridisation. There are also a number of approaches to measuring the background.

Most image processing software makes a *local estimate of the background* by identifying some number of pixels surrounding each spot and using those to calculate an average or median background level that can be subtracted from each pixel in the spot. This has the advantage that it treats each spot separately, allowing variations in the background to be individually estimated. One possible disadvantage to this approach is that if the pixels selected contain portions of either the target spots or those surrounding it, one may overestimate the background significantly. Consequently, some image processing software allows the estimation of *global background* – a single measurement that is used for the entire slide. Since only one measurement will be used, this generally requires users to select a large representative area of the slide that is devoid of features. The disadvantage of this, of course, is that it fails to account for local variations in the background fluorescence of the substrate, and as a result it may provide either an over- or underestimate of the appropriate background.

If there is significant non-specific hybridisation to the surface of the microarray, both local and global methods may overestimate the background. Many people have reported the presence of 'negative spots' or 'black holes' in arrays where there has been incomplete blocking of the (glass) surface prior to hybridisation. This phenomenon occurs when the labelled extract binds with higher efficiency to the surface itself than it does to the corresponding features on the array, with the result that the features themselves serve as blocking agents reducing non-specific hybridisation. To deal with this, some image processing software allows users to specify negative hybridisation control spots. These typically consist of exogenous DNA selected to have as little homology as possible with the species of interest so that all we measure is the fluorescence of the microarray surface under the spot and any non-specific hybridisation to DNA. This has the advantage that it measures precisely what we want to subtract from each spot, but it does not take into account local variation in the slide surface fluorescence nor does it precisely account for sequence-mediated differences in non-specific hybridisation.

3.2.2.4 REPORTING EXPRESSION WITH AFFYMETRIX GENECHIPS^{TM}

As discussed previously, most image processing software reports a variety of measured parameters for each feature on the array, including mean, median and total intensities, spot areas, background estimates, and statistical properties of the feature and background pixel intensity distributions. Affymetrix GeneChips^{TM} software is a notable exception. The Affymetrix approach to microarray analysis uses features consisting of 25 base oligonucleotides selected from each parent gene's sequence. For each gene, 11 to 20 paired sets of 'perfect match' (PM) and 'mismatch' (MM) are synthesised, such that PM/MM pairs are in adjacent positions on the array. The PM feature is a perfect match to the sequence while the MM has a single mismatch nucleotide in the middle of the strand (position 13 of 25). The MM features are designed to provide a feature-by-feature estimate of the non-specific hybridisation that one expects for the PM features and play a role equivalent to the background measures used in most other systems. As a measure of the expression of the gene, the Affymetrix image processing software reports an average difference (AD),

$$AD \sim \frac{\sum_{k=1}^{N} (PM_k^{\text{intensity}} - MM_k^{\text{intensity}})}{N}$$

where N is the number of feature pairs for that particular gene. It is this average difference that is used in subsequent calculations as a background-subtracted measure of gene expression.

One potential problem with this approach is that it can result in negative values for the measured expression levels. Recent versions of the Affymetrix expression reporting software use a scheme that decreases the relative contribution of feature pairs with large mismatch values,

$$\text{Signal} \sim \frac{\sum_{k=1}^{N} w_k * (PM_k^{\text{intensity}} - MM_k^{\text{intensity}})}{\sum_{k=1}^{N} w_k}$$

where w_k are the weights assigned to each feature pair. A simple approach adopted by some users of Affymetrix GeneChips^{TM} prior to the release of the most recent software system was to use a binary scoring scheme,

$$w_k = \begin{cases} 1 & PM > MM \\ 0 & PM < MM \end{cases}$$

which eliminates features with greater hybridisation to the mismatch probe and guarantees that the average difference value be positive. This has been shown to provide much more reproducible measures of expression (Irby *et al.*, 2002; Li and Wong, 2001).

3.2.2.5 *EXPRESSION RATIOS: THE STARTING POINT FOR SAMPLE COMPARISON*

Regardless of how we choose to report expression for each gene in an individual RNA sample, most analyses focus on differences in gene expression, which are usually reported as a background-subtracted ratio of expression levels between a selected query sample and a reference sample. Typically, this is done by taking ratios of the measured expression between two physical states,

$$T_k = \frac{R_k}{G_k}$$

for each gene k on the array, where we designate the query and reference samples R and G, respectively (as one might choose for a two-colour DNA microarray where the Cy3 sample is usually represented as green and the Cy5 channel is represented as red). We will use this notation for what we present here, but it should be noted that it is not specific to two-colour DNA microarrays. The same formalism could be applied to separate measurements on a single-colour array, where the R and G measurements represent those on independent arrays, or in the Affymetrix GeneChip^{TM} system, where R and G represent the derived gene expression values from two arrays being compared.

As noted previously, there are a variety of methods for representing the measured expression for each feature on the array. Consequently, there are a variety of methods for calculating the expression ratio using the median, mean and integrated expression measures. If we choose to measure expression using the median pixel value for each feature, then the median expression ratio for a given feature is

$$T_{median} = \frac{R^{feature}_{median} - R^{background}_{median}}{G^{feature}_{median} - G^{background}_{median}}$$

where $R^{feature}_{median}$ and $R^{background}_{median}$ are the median intensity values measured for pixels identified in the feature and background, respectively. In a similar fashion, if $A_{feature}$ is the area of a given feature, we can define the background-subtracted ratio of integrated intensities as

$$T_{\text{integrated}} = \frac{\sum\limits_{i \in (\text{spot pixels})} (R_i^{\text{feature}} - R_{\text{mean}}^{\text{background}})}{\sum\limits_{i \in (\text{spot pixels})} (G_i^{\text{feature}} - G_{\text{mean}}^{\text{background}})} = \frac{R_{\text{integrated}}^{\text{feature}} - (R_{\text{mean}}^{\text{background}} \times A_{\text{feature}})}{G_{\text{integrated}}^{\text{feature}} - (G_{\text{mean}}^{\text{background}} \times A_{\text{feature}})}$$

It should be noted that the ratio of the integrated background-corrected spot intensities is equivalent to the ratio of the mean intensities,

$$T_{\text{mean}} = \frac{\left[\sum\limits_{i \in (\text{spot pixels})} (R_i^{\text{feature}} - R_{\text{mean}}^{\text{background}}) \right] \Big/ A_{\text{feature}}}{\left[\sum\limits_{i \in (\text{spot pixels})} (G_i^{\text{feature}} - G_{\text{mean}}^{\text{background}}) \right] \Big/ A_{\text{feature}}}$$

$$= \frac{\sum\limits_{i \in (\text{spot pixels})} (R_i^{\text{feature}} - R_{\text{mean}}^{\text{background}})}{\sum\limits_{i \in (\text{spot pixels})} (G_i^{\text{feature}} - G_{\text{mean}}^{\text{background}})} = T_{\text{integrated}}$$

3.2.2.6 TRANSFORMATIONS OF THE EXPRESSION RATIO

Ratios are useful because they allow us to measure expression differences in an intuitive way, as genes that do not change in their expression level have a ratio of 1. However, ratios are troublesome because they treat up- and down-regulated genes differently. Genes up-regulated by a factor of 2 have an expression ratio of 2, while those down-regulated by the same factor have an expression ratio of $\frac{1}{2}$ (0.5). As a result, down-regulated genes are compressed between 1 and 0 while up-regulated genes expand to cover the region between 1 and positive infinity.

To rectify this problem, the ratio is often transformed to provide a similar representation for differentially expressed genes regardless of whether they are up- or down-regulated. The *reciprocal* or *inverse transformation* converts the ratio to a *fold change*. For genes with expression ratios less than 1, one takes −1 times the reciprocal of the expression ratio, or

$$(\text{fold change})_k = \begin{cases} T_k & T_k \geq 1 \\ -\dfrac{1}{T_k} & T_k < 1 \end{cases}$$

for the *k*th gene on the array. The advantage of this approach is obvious: genes whose expression increases by a factor of 2 have a fold change of +2 while those that decrease by the same factor have a fold change of −2, providing an intuitive interpretation for changes. The disadvantage, however, is that the fold change is discontinuous between −1 and +1 and as a result is not amenable to most mathematical analyses.

A better alternative is to apply a *logarithmic transformation*, generally using the logarithm base 2. The advantage of this transformation is that it produces a continuous spectrum of values for differentially expressed genes while treating up- and down-regulated genes equivalently. Recall that logarithms treat numbers and their reciprocals symmetrically: $\log_2(1) = 0$, $\log_2(2) = 1$, $\log_2(\frac{1}{2}) = -1$, $\log_2(4) = 2$, and $\log_2(\frac{1}{4}) = -2$. If we take the logarithm of the expression ratios, these will also be treated symmetrically so that a gene up-regulated by a factor of 2 has a $\log_2(\text{ratio})$ of 1, a gene down-regulated by a factor of 2 has a $\log_2(\text{ratio})$ of -1, and a gene expressed at a constant level (with a ratio of 1) has a $\log_2(\text{ratio})$ equal to 0. The relationship between the ratio and the $\log_2(\text{ratio})$ can be seen in Figure 3.3.

It should be noted that there are disadvantages to using expression ratios, or the transformations derived from them, for data analysis. While ratios can help to reveal some patterns in the data, they remove all information about absolute gene expression levels. A variety of parameters depend on the measured intensity, including the confidence limits one places on any microarray measure-

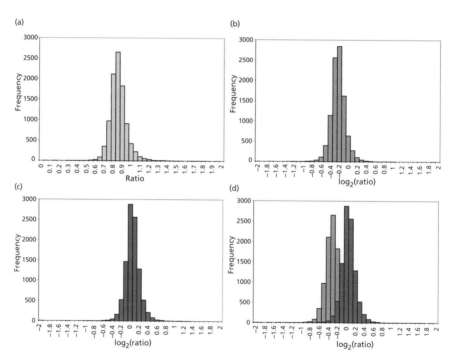

Figure 3.3 Gene expression ratios and the effect of normalisation. (a) Comparisons of gene expression patterns are generally represented by ratios comparing measured hybridisation intensities. (b) Ratios, however, have the disadvantage of compressing measurements below 1 and expanding them above 1; consequently, the logarithm of the ratio is often used. In a comparison of related samples, one expects the ratio to be 1, or the corresponding $\log_2(\text{ratio})$ to be 0. (c) and (d) Normalisation adjusts the ratios appropriately to remove experimental biases that might be present, allowing differential expression to be more accurately assessed.

ment. While most of the techniques developed for analysis of microarray data use ratios, many of them can be adapted for use with measured intensities, although none is yet in widespread use.

3.2.2.7 SITUATIONS WHERE EXPRESSION DOES NOT CORRELATE WITH SPOT INTENSITY

We must also remember that all the measurements we make are estimates of the expression level for each gene and that there are a number of factors that can distort the relationship between expression and the measurements we make. We have already discussed a number of possible sources of deviation, including poor estimation of the median or total intensities, pixel saturation, and background estimation, but others exist. If the density of fluors in the labelled, hybridised molecules is high enough, interaction between the dye molecules can quench fluorescence. Similarly, poor labelling or hybridisation can result in signals too faint to allow detection of certain expressed genes.

There are a number of other factors that may contribute to incorrect estimates of hybridisation. If there is significant cross-hybridisation or non-specific hybridisation between arrayed features, any measurements may be incorrect. If we use cDNA clones to construct the arrays, chimeric or misidentified clones may lead us to erroneous conclusions. PCR amplicons or oligonucleotides may be contaminated with other DNAs and may not accurately bind only the gene of interest, and oligonucleotide sequences may be incorrectly synthesised. Hybridisation between alternative splice forms and members of gene families may also cause us to overestimate fluorescence and therefore expression, even in the most stringent hybridisation assays.

Finally, one must remember that microarray assays allow us to measure only relative RNA levels within samples, not expression itself. Sample handling can clearly influence what we measure and may lead to conclusions that reflect less about the biology of the situation we are studying and more about our laboratory practices. Manipulation of tissue samples, hypoxia, cold or heat shock, and cell death can cause genes to be expressed that are not normally present in the tissues we want to study, and RNA degradation can skew representation, as some RNAs are more stable than others. The best protection against any of these artefacts is to institute validated laboratory protocols selected to provide the best quality RNA, labelling, hybridisation and scanning. No image processing or other software can save bad quality data, and the best, most reproducible results are invariably obtained from the cleanest hybridisations.

3.3 Normalisation

While ratios or their transformations are a good means of comparing levels of gene expression between a query and a reference sample, before differentially expressed genes can be reliably identified one must first have an accurate method for comparing the measured expression levels between states. There are a variety of reasons why the raw measurements of gene expression for two samples may not be directly comparable: the quantity of starting RNA may not be equal for each of the samples, there may be differences in labelling and detection efficiencies for the fluorescent labels, and there may be additional systematic effects that can skew the measured expression levels and the derived expression ratios. Consider a *self–self comparison* in which the same sample is compared with itself using independent labelling reactions. Here, one expects the measured $\log_2$(ratio) to be 0 for each gene (equivalent to the ratio being 1) since there should be no difference between the representation of genes in the starting RNA. However, typically these are found to be distributed with a non-zero mean and standard deviation, indicating that there is a bias to one sample or the other, as well as an inherent uncertainty and an associated variation in each measurement.

Normalisation is any data transformation that adjusts for these effects and allows the data from two samples to be appropriately compared. In addition, there are a number of other data transformations that we may want to apply before proceeding with any more complex analyses. For example, we may want to identify and eliminate questionable and low quality data or to average across replicate measurements. Finally, we may want to perform additional analyses, such as searching for differentially expressed genes in the datasets we generate.

Normalisation scales one or both of the measured expression levels for each gene to make them equivalent, and consequently the expression ratios derived from them. This is done in such a way that the average expression levels are made equivalent for the two samples being compared. The results of normalising microarray data can be seen in Figures 3.3 and 3.4. Because normalisation changes the data, one has to understand both the basic principles that each technique employs and how it changes the data. Further, all normalisation strategies are based on some underlying assumptions regarding the data and the experimental design, and consequently the normalisation approach that is used must be appropriate to the particular experiment.

Normalisation approaches typically use either the complete set of arrayed genes or a *control set*, generally either a set of *housekeeping genes* or a set of *exogenous spiked-in controls*. Housekeeping genes are those that, in the systems under study, are assumed either singly or collectively not to change in expression level. Exogenous controls are genes from a species other than that under study, generally selected not to cross-hybridise with other genes included in the array (e.g. *Arabidopsis* genes for photosynthesis included in mammalian microarray experiments), and for which a stable source of RNA is available.

The advantage to using a control set is that the only assumption required is that those genes will be detected at constant levels in all of the samples being compared. However, it requires careful quantitation of the initial RNA that is used in each labelling reaction, it ignores much of the data, and it often fails to account for any variation-dependent expression level. If the control set represents housekeeping genes, these must, in fact, not change in expression in the conditions being surveyed, and these sets can easily be biased by incorrect as-

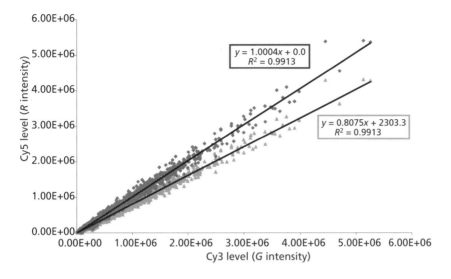

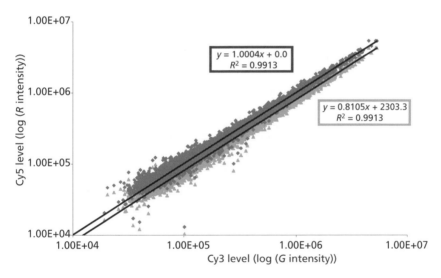

Figure 3.4 Scatterplots of measured intensities, or their logarithms, can be used to visualise the effects of normalisation on expression data. Normalisation shifts the average ratio to 1, resulting in an adjustment of the best-fit slope to 1.

sumptions. Although it is commonly accepted that there is a set of 'housekeeping genes' of relatively invariant expression, there is accumulating evidence to suggest that many of these genes change in expression under some circumstances (Lee *et al.*, 2002; Thellin *et al.*, 1999). This makes it important to use a large number of control genes as a normalisation set. For strategies using the

Table 3.1 Techniques for calculating normalisation factors, based on different properties of the data.

Normalisation set	Normalisation type/based on:
All/Percentile of all features	e.g. Total intensity, mean, median intensity, linear regression, Chen's ratio statistics
Subset of genes, based on: (a) Source	
– endogenous to sample	e.g. Housekeeping genes
– exogenous to sample	e.g. Spiked-in controls
(b) Signal intensity	e.g. Lowess
(c) Location of feature on array	e.g. Local normalisation

entire dataset, RNA quantitation is somewhat less important and the statistical support for any correction is generally better, but these lack an independent confirmation and can be skewed if the majority of genes on the array are differentially expressed. A third alternative is to choose an unbiased set of genes selected by *rank ordering* the genes or signal from each spot, placing them in order based on expression level and using only those within a fixed window centred within the dataset (e.g. those between the 30th and 70th percentile) or those within a fixed number of standard deviations of the mean (Schadt *et al.*, 2001; Tseng *et al.*, 2001). In fact, no single approach can be applied in all circumstances, and that used must be selected with the experimental system being examined in mind.

Once a normalisation gene set has been selected, a *normalisation factor* (sometimes referred to as a *scaling factor*) is calculated for each gene and used to adjust the data to compensate for the observed experimental variability and to 'balance' the fluorescence signals from the labelled extracts being compared. There are a number of techniques that can be used to calculate normalisation factors (Table 3.1), and we examine some of these below. The measured intensities referred to in the sections that follow assume that the values representing the expression of each spot have been corrected to account for differences in the background signal.

3.3.1 Total intensity normalisation

Of all the techniques that have been used for microarray data analysis, total intensity normalisation is probably the easiest to understand. Total intensity normalisation relies on two assumptions. First, we assume that the total quantity of starting mRNA is the same for both samples being analysed and hence, that we have approximately the same number of molecules in each. While particular genes may be up-regulated in the sample, others must be down-regulated to compensate; this is equivalent to assuming that cells have a certain RNA synthesis capacity. Second, we must assume that the genes represented on the array, or the subset we choose for normalisation, are not biased to significantly over-represent genes expected to be differentially expressed between the samples we are analysing. Consequently, for the hundreds or thousands of genes in the array, changes in expression level between the labelled extracts under comparison should balance out, so that the total quantity of RNA hybridising to the array from each sample is the same. If this is true, then the hybridisation intensities for both of the samples being compared, summed over all of the features in the normalisation set, should be equal.

Under this assumption, a normalisation factor can be calculated and used to rescale the intensity for each gene in the array. One calculates a normalisation factor by summing the measured intensities in both channels

$$N_{total} = \frac{\sum_{k=1}^{Narray} R_k}{\sum_{k=1}^{Narray} G_k}$$

where G_k and R_k are, respectively, the measured intensities for the kth array feature (such as the Cy3 and Cy5 intensities in a two-colour microarray assay) and N_{array} is the total number of features represented in the microarray; alternatively, the summation runs over the subset of features selected as normalisation standards. The intensities are then rescaled such that

$$G'_k = N_{total}G_k \quad \text{and} \quad R'_k = R_k$$

and the normalised expression ratio for each feature becomes

$$T'_k = \frac{R'_k}{G'_k} = \frac{1}{N_{total}}\frac{R_k}{G_k}$$

which adjusts the ratio such that the mean ratio is equal to 1; this transformation is equivalent to subtracting a constant from the logarithm of the expression ratio,

$$\log_2(T'_k) = \log_2(T_k) - \log_2(N_{total})$$

This sort of global normalisation is an example of a *single-parameter linear normalisation*: one that uses a single factor to scale the data. It should be noted that what we have described here is not the only acceptable approach. Obviously, one could scale the intensities so that the mean or median intensities are the same across arrays, or that any of these are set equal to a fixed constant.

3.3.2 Mean log centring

One can easily see that total intensity normalisation approach is related to mean log-centring normalisation, in which one relies on the assumption that the mean $\log_2(\text{ratio})$ should be equal to 0. In this case, we calculate a normalisation constant that is equal to the mean $\log_2(\text{ratio})$ for all of the features in the normalisation set,

$$N_{\text{mlc}} = \frac{1}{N_{\text{array}}} \sum_{k=1}^{\text{Narray}} \log_2(T_k) = \frac{1}{N_{\text{array}}} \sum_{k=1}^{\text{Narray}} \log_2\left(\frac{R_k}{G_k}\right)$$

Using this, the $\log_2(\text{ratio})$ for each feature is then scaled

$$\log_2(T_k') = \log_2(T_k) - N_{\text{mlc}}$$

thus guaranteeing that the average $\log_2(\text{ratio})$ is 0. Of course, this is equivalent to adjusting the intensities

$$G_k' = 2^{N_{\text{mlc}}} G_k \quad \text{and} \quad R_k' = R_k$$

or the ratio

$$T_k' = \frac{R_k'}{G_k'} = \frac{1}{2^{N_{\text{mlc}}}} \frac{R_k}{G_k}$$

in a fashion similar to that described previously.

One disadvantage of this approach is that it is sensitive to outlying, differentially expressed genes. If there are a number of genes that are significantly up-regulated, for example, one will overestimate N, resulting in an overcorrection of the expression ratios. A variation of this approach is *iterative mean log centring*. First, ratios are adjusted such that the mean $\log_2(\text{ratio})$ for the entire collection of genes is set to 0 (or a corresponding average ratio of 1) as described above. Outliers are then identified and excluded, a new mean is calculated for the remaining data, and this is used to normalise all of the data; this process is repeated until convergence.

3.3.3 Linear regression

An alternative approach is to use linear regression analysis. For closely related samples, one would expect many of the genes to be expressed at nearly constant levels. In a scatterplot of intensities (or their logarithms) from the two samples being compared, these genes would cluster along a straight line, the slope of which would be 1 if the labelling and detection efficiencies were the same for both samples. Normalisation of these data is equivalent to calculating the best-fit slope using regression techniques (Chatterjee and Price, 1991) and adjusting the intensities so that the calculated slope is 1.

Regression analysis assumes that the measured intensity for each gene in the first sample, R_k, is an approximate linear function of the intensity in the second channel, G_k, so one can model the relationship as

$$R_k = \beta_0 + \beta_1 G_k + u_k$$

where β_0 and β_1 are constants representing the intercept and slope, respectively, and u_k is random, independently distributed noise in the measured signal with mean 0 and a common variance. The best-fit estimate of the slope, b_1, is given by

$$b_1 = \frac{\sum_{k=1}^{n}(R_k - \overline{R})(G_k - \overline{G})}{\sum_{k=1}^{n}(G_k - \overline{G})^2}$$

where $\overline{G}$ and $\overline{R}$ are, respectively, the mean of all the values in the two samples. The estimate of the intercept, b_0, is then simply

$$b_0 = \overline{R} - b_1 \overline{G}$$

The measured intensities for each feature, k, of the array can then be rescaled such that

$$G'_k = \left[\frac{G_k - b_0}{b_1}\right] \quad \text{and} \quad R'_k = R_k$$

which again assures that the appropriate ratios are scaled, on average, to 1.

3.3.4 Chen's ratio statistics

Another approach is the ratio statistics method developed by Chen *et al.* (1997). They assume that while individual genes may be up- or down-regulated, in closely related cells the total amount of mRNA is the same for a set of 'housekeeping genes' in the extracts being compared. For these, they assume that there exists a distribution of expression levels, with a common mean μ and standard deviation σ independent of the sample. Under this assumption, they develop an approximate probability density for the ratio $T_k = R_k/G_k$ and describe an iterative process that normalises the mean expression ratio to 1 and calculates confidence limits that can be used to identify differentially expressed genes. Chen and collaborators have since extended this approach to use the entire available dataset (personal communication).

3.3.5 Lowess normalisation

It has been noted in a number of reports (see Yang *et al.*, 2001) that the $\log_2$(ratio) values often have a systematic dependence on intensity, most often observed as a deviation from 0 for low intensity spots. Locally weighted linear regression, *Lowess* (Cleveland, 1979), has been proposed as a normalisation method for microarray assays that can remove intensity-dependent dye-specific effects in the $\log_2(\text{ratio})$ values (Yang *et al.*, 2002a,b).

The starting point for this analysis is plotting the measured $\log_2(R/G)$ ratios for each array feature as a function of the $\log_{10}(R \star G)$ product intensities. The resulting 'R–I plot' (for *ratio–intensity*) can reveal intensity-specific artefacts in the measurement of the ratio, which tend to occur most notably for weakly fluorescing arrayed features.

Our expectation, similarly to the other approaches described here, is that the mean $\log_2(R/G)$ ratio should be 0 for the arrayed features, independent of intensity. A representative R–I plot for a self–self hybridisation, in which the same RNA sample was labelled with both Cy3 and Cy5 and hybridised to a 19,200 feature cDNA array, is shown in Figure 3.5. In this case, we expect to see absolutely no differential expression and consequently all $\log_2(\text{ratio})$ measures should be (on average) 0. However, inspection of this R–I plot clearly shows a slight upward curvature at both high and low intensities, as well as an increased spread in the distribution of $\log_2(\text{ratio})$ values at low intensities.

Lowess detects such deviations from the expected behaviour and corrects them by performing a local weighted linear regression for each data point in the R–I plot and subtracting the calculated best-fit average $\log_2(\text{ratio})$ from the experimentally observed ratio for each point as a function of the $\log_2(\text{intensity})$.

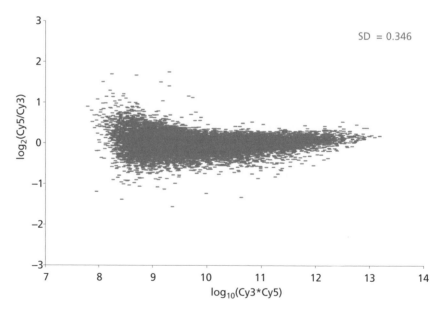

Figure 3.5 A widely used diagnostic for gene expression measurements is the R–I plot, in which the logarithm of the ratio of measured intensities is plotted as a function of the logarithm of their product. R–I plots clearly display systematic variation of the ratio as a function of the intensity. Here, one can see an increase in the average ratio at high and low intensities, as well as the general increase in the spread of the measured ratio at low intensities. SD, standard deviation. (Yang *et al.*, 2002a)

The process is conceptually similar to performing a local, weighted mean log centring.

For each point in the R–I plot, Lowess uses a weight function that de-emphasises the contribution of data from array features that are far from our point of interest, so that the farther the distance between the query data and its neighbouring data, the less the neighbouring data will affect the fit and the re-sulting regression function. In practice, this is similar to calculating a running correction, using only the data in a local neighbourhood for the calculation, and then weighting even those. There are many weight functions that can be applied, but the most commonly used is the *tri-cube weight function*,

$$w(u) = \begin{cases} (1 - |u^3|)^3 & |u| < 1 \\ 0 & |u| > 1 \end{cases}$$

where u is the distance from a particular data point to those in its neighbour-hood. If we set $x_k = \log_{10}(R_k * G_k)$ and $y_k = \log_2(R_k/G_k)$, Lowess then performs a rather straightforward weighted linear regression to produce an estimate, $y(x_k)$, of the dependence of the $\log_2$(ratio) on the $\log_{10}$(intensity). This function can then be used, point by point, to correct the measured $\log_2$(ratio) values, so that

$$\log_2(T_k') = \log_2(T_k) - y(x_k) = \log_2(T_k) - \log_2(2^{y(x_k)})$$

or

$$\log_2(T_k') = \log_2\left(T_k * \frac{1}{2^{y(x_k)}}\right) = \log_2\left(\frac{R_k}{G_k} * \frac{1}{2^{y(x_k)}}\right)$$

As with the other normalisation methods, we can make this equivalent to a transformation on the intensities,

$$G_k' = G_k * 2^{y(x_k)} \quad \text{and} \quad R_k' = R_k$$

The result of applying a Lowess transformation can be seen in Figure 3.6.

An alternative to applying Lowess to the R–I plot would be to apply it directly to the log-transformed intensities. One would simply plot $y = \log_2(R_k)$ against $x = \log_2(G_k)$ for each feature, apply Lowess to approximate the depend-ence of $\log_2(R_k)$ on $\log_2(G_k)$, and use the calculated best-fit curve to calculate a normalisation factor that would adjust the best-fit slope to 1.

3.3.6 Global vs. local normalisation

Most normalisation algorithms can be applied either globally to the entire dataset or locally to subsets of the data, such as individual *subgrids* (i.e. an area of the array where all the spots were deposited by a single spotting pen, some-

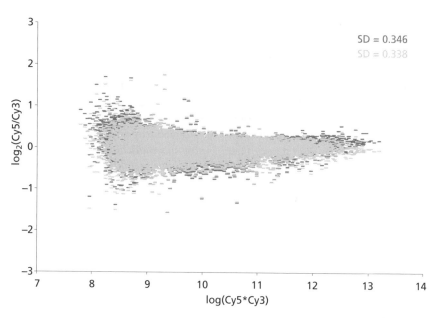

Figure 3.6 The effect of Lowess normalisation. Lowess can be used to remove the systematic variation in the mean gene expression ratio as a function of intensity. The normalised data, shown in light grey, are symmetrically distributed around a mean $\log_2$(ratio) of 0, unlike the un-normalised data shown in dark grey. (Yang *et al.*, 2002a)

times also referred to as a *pen group*). Applying these algorithms to a single sub-grid can help to correct for local systematic effects due to a variety of sources, including variability between the spotting pens used to make the array, variability in the slide surface, and slight local differences in hybridisation conditions across the array.

However, if one is to apply a particular normalisation algorithm locally, all of the conditions and assumptions that underlie the validity of the approach must be satisfied. If the algorithm uses exogenous spiked-in controls, or a set of selected housekeeping genes, then all of these control features must be present in each subgrid. If, instead, the approach chosen uses all of the data, one must be sure both that the features in any subgrid were not selected to preferentially represent differentially expressed genes and also that a sufficiently large number of features is included in each subgrid for the approach to be valid.

3.4 Data filtering

Normalisation is only one of many transformations that can be applied to microarray datasets. The goal of most other transformations is to filter the dataset to reduce its complexity and increase its overall quality. Many are de-

signed to flag questionable and low quality data, while others are used to identify differentially expressed genes or to enhance a particular feature of the data.
Again, which methods you choose to apply will depend on your particular experimental design and on the biological questions you are seeking to address.

3.4.1 Filtering low intensity data

One of the most widely used data filters eliminates those array features that have
hybridisation intensities only slightly above the measured background. The justification for this approach is simple: those spots with intensities near the background are the most imprecisely measured and therefore most likely to be of
poor or questionable quality. Typically, one calculates the local background for
each array feature and its standard deviation in each channel, $\sigma(G^{\text{background}})$ and
$\sigma(R^{\text{background}})$. If we are using background-subtracted median values for the
hybridisation intensities, those less than two times the standard deviation of
the background are flagged and eliminated, so that we use only spots where the
measured signals meet the criteria that

$$G^{\text{spot}}_{\text{median}} > 2 * \sigma(G^{\text{background}}) \quad \text{and} \quad R^{\text{spot}}_{\text{median}} > 2 * \sigma(R^{\text{background}})$$

and similarly for the mean. If instead we use the integrated intensities, we must
take into account the measured areas for each spot, so that useful spots are those
where

$$G^{\text{spot}}_{\text{integrated}} > 2 * \sigma(G^{\text{background}}) * A_{\text{spot}}$$

and

$$R^{\text{spot}}_{\text{integrated}} > 2 * \sigma(R^{\text{background}}) * A_{\text{spot}}$$

As an alternative, one can use a percentage-based cut-off in which spots are
eliminated if they have intensities less than some arbitrary fraction of the mean
or median background. As another alternative, some groups use fixed minimum intensities to identify and eliminate array features they feel are questionable, or eliminate array features falling in the lowest fifth or tenth percentile of
intensity values in each channel. While these alternative approaches do not have
any particular statistical justification, they can help to improve the general quality of the data from each hybridisation and the overall quality of any subsequent
analysis.

3.4.2 Setting floors and ceilings

Another approach that is sometimes used to adjust the low and high intensity
data is to set a *floor* and *ceiling*, minimum and maximum acceptable values for
intensities. Those data below the floor value are adjusted upwards and set equal

to the floor, and those above the ceiling are set equal to the ceiling value. This has the effect of reducing the fold change value for genes expressed at low and high levels. Such setting of floors is useful for analysis of genes with low expression levels, as these measures are most likely to be inaccurate. In particular, floored values can be used to obtain meaningful ratios for genes that are turned on or off between conditions. It is good practice to flag genes that have been adjusted to the ceiling or floor as the expression ratios computed for these are artificial constructs that do not reflect the experimental measurements. One way to select floor and ceiling values is to find the measured intensity values at which the relationship between expression in the samples being compared begins to deviate from linearity.

3.4.3 Use of replicate data

Replication is essential for identifying and reducing the effect of variability in any experimental assay, and microarray analysis is no exception (the generation and use of replicate data is also discussed in Sections 2.2.3 and 4.2.1). Generally, we divide replicates into two broad classes (Figure 2.2). Biological replicates use independently derived RNA from distinct biological sources to provide an assessment of both the variability in the assay and the inherent biological variability in the system under study. Biological replicates allow commonly expressed genes to be identified, as well as those that are distinct to the particular biological sample. Technical replicates, on the other hand, provide only information on the variability of the assay. These include replicate features within a single microarray, multiple independent features for a particular gene within an array, or replicated hybridisations for a particular sample. The specific approach used will of course depend on the biological questions being asked and the experimental design used for a particular study.

3.4.4 Experimental design strategies

Many of the techniques that have been developed for filtering data depend on the design that is used for a particular experiment. Most microarray experiments use a *reference design* in which all of the biological samples analysed are compared with a single common reference sample. One advantage of this approach is that interpretation of the data and comparison between samples is rather straightforward: one assumes that genes with similar patterns of expression relative to the reference would in fact exhibit similarities if directly compared. Reference designs are also easily extensible, provided that additional reference RNA is available. However, one should note that in two-colour hybridisation experiments using a reference design, the common reference sample, which is generally the least interesting of all those being assayed, is the one sample on which we collect the greatest quantity of data on gene expression levels.

As an alternative, Kerr and Churchill (2001) have proposed using reference-less *loop designs* in which labelled extracts are serially compared with each other rather than to a common reference (also discussed in Section 2.2.4). The advantage of such an approach in two-colour array assays is that each experimental sample is queried more often than would be possible with a reference design using the same number of experimental assays. This allows the actual expression for each gene in each condition to be estimated with greater accuracy than would otherwise be possible. There are, however, two potential drawbacks to loop designs. First, all the experimental samples must be collected prospectively and in great enough quantities that all of the required hybridisations can be carried out. If there is insufficient RNA from a particular sample, or if the RNA from a sample does not label or hybridise well, then the loop will contain missing data, and estimating expression may be difficult or even impossible. While one might be able to replace some hybridisations, this would also require RNA from one or more of the other samples being assayed. Second, loop designs are not easily extensible, and adding new samples to an analysis is often impossible. For these reasons, loop designs are often impossible to implement for clinical studies using samples from human patients, but often perform well in laboratory and model systems.

However, whether one uses a reference or a loop design, it is clear that replicate hybridisations are essential for improving the accuracy of the measurements made. For two-colour arrays, it is also advisable to use a dye-swap (or dye-flip) approach (see Section 2.2.6) in which replicates are performed with the fluorescent labels for the two samples exchanged between hybridisations. The advantage of this is that dye-specific effects can more easily be detected and, if necessary, eliminated.

3.4.5 Replicate filtering

Replicate filtering can allow the identification of questionable or low quality spots by highlighting inconsistencies in the hybridisation measures from replicates. For convenience, we will focus our discussion on the analysis of data from a pair of dye-swap technical replicates, but this approach can be generalised for any replication strategy.

First assume that we are comparing two samples, A and B. In our first experiment, we will label A with Cy5 (red, R) and B with Cy3 (green, G). For the ith gene arrayed as a feature, we measure the expression ratio

$$T_{1i} = \frac{R_{1i}}{G_{1i}} = \frac{A_{1i}}{B_{1i}}$$

If we repeat this measurement for a second technical replicate, swapping the dyes, the ith gene gives us the expression ratio

$$T_{2i} = \frac{R_{2i}}{G_{2i}} = \frac{B_{2i}}{A_{2i}}$$

As we are making two comparisons between identical samples, we expect

$$(T_{1i} * T_{2i}) = \left(\frac{A_{1i}}{B_{1i}} * \frac{B_{2i}}{A_{2i}} \right) = 1$$

or equivalently,

$$\log_2(T_{1i} * T_{2i}) = \log_2\left(\frac{A_{1i}}{B_{1i}} * \frac{B_{2i}}{A_{2i}} \right) = 0$$

where A_{1i} and A_{2i} are replicate measurements of the expression of the ith gene in sample A and B_{1i} and B_{2i} are replicate measurements of the ith gene in sample B.

Experimental variation in each of the measurements will lead to a distribution of the measured values for the log of the product ratios, ($T_{1i} * T_{2i}$). One can then calculate the mean and standard deviation for the arrayed genes. Those that deviate significantly from the expected value of 0 are likely to contain one or more poor quality features. Although one cannot determine *a priori* which of the replicates is likely to be in error, visual inspection may allow the 'bad' spot to be identified and removed before further analysis. Alternatively, and more practically for large experiments, one can simply eliminate questionable spots. An example of the use of such replicate filtering is shown in Figure 3.7, where $\log_2$(ratio) values from replicate experiments in which two cell lines, one derived from colon and the second from pancreas, were compared on a 19,200 feature cDNA array. In this case, one expects the replicate ratios to be linearly related to each other. The correlation coefficient increased from $r = 0.84$ (dark grey; dashed line) to $r = 0.96$ (light grey; solid line) after 1.7% of outliers were filtered out using the method described above. The substantial increase in the correlation coefficient suggests such a filtering procedure efficiently removes features with at least one unreliable $\log_2$(ratio) value. Obviously, a similar approach can be used to identify and eliminate questionable spots using within-array replicates.

3.4.6 Averaging replicate data

Many users also want to average their replicates to produce a single consensus measurement and thereby reduce the complexity of the final dataset. If we focus on the $\log_2$(ratio) for the ith array feature, then what we are seeking is a constant c which we can use to adjust each of the individual measurements such that

$$\log_2\left(\frac{A_{1i}}{B_{1i}} \right) + c_i = \log_2\left(\frac{A_{2i}}{B_{2i}} \right) - c_i$$

This constraint equation can be easily solved to yield a value for the constant c_i:

$$c_i = \frac{1}{2}\log_2\left(\frac{A_{2i}}{B_{2i}} \frac{B_{1i}}{A_{1i}} \right) = \log_2\left(\sqrt{\frac{A_{2i}}{B_{2i}} \frac{B_{1i}}{A_{1i}}} \right)$$

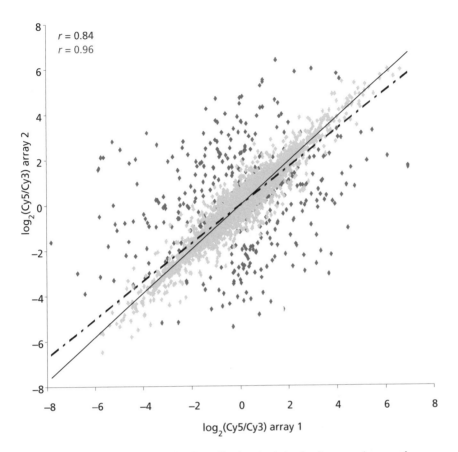

Figure 3.7 An example of the use of replicate filtering. Analysis of replicates can improve the accuracy of the ratio measures by removing outlying, low confidence data, as can be seen in this comparison of measured $\log_2$(ratio) values before (dark grey; dashed line) and after (light grey; solid line) filtering. (Yang *et al.*, 2002a)

If we use this to correct either of our two measurements, the result is equivalent to taking the geometric mean of our measurements, or

$$\log_2\left(\frac{\overline{A}_i}{\overline{B}_i}\right) = \log_2\left(\sqrt{\frac{A_{2i}}{B_{2i}}\frac{A_{1i}}{B_{1i}}}\right)$$

where the average measurements for expression in each of the samples is given by, respectively,

$$\overline{A}_i = \sqrt{A_{1i}A_{2i}} \quad \text{and} \quad \overline{B}_i = \sqrt{B_{1i}B_{2i}}$$

$\overline{A}_i$ and $\overline{B}_i$ can then be used as to create an R–I plot, with $\log_2(\overline{A}_i/\overline{B}_i)$ plotted as a function of the $\log_{10}(\overline{A}_i \star \overline{B}_i)$ product intensities for each arrayed feature, or for any other application.

3.5 Identification of differentially expressed genes

In many experiments, one of the goals is to find genes that are differentially expressed between two states. After normalising, filtering and averaging the data, one can then identify genes with expression ratios that are significantly different from 1, or equivalently, those with values for the $\log_2(\text{ratio})$ that differ from 0. Many early published microarray studies used an arbitrary cut-off of two-fold up- or down-regulation as significant, or equivalently, $\log_2(\text{ratio})$ values greater than 1 or less than -1.

Such arbitrary, fixed cut-offs may or may not be supported by the data or by the underlying biology. For example, we know that the expression of some genes fluctuates a great deal more than others, as well illustrated in the analysis by Hughes *et al.* (2000a,b), in which replicate hybridisations were examined to build gene-specific error models. In general the genes whose expression is most variable are those in which expression is stress induced, modulated by the immune system or hormonally regulated (Pritchard *et al.*, 2001). In a well-conducted experiment comparing closely related samples, the distribution of $\log_2(\text{ratio})$ values might be such that there is very little spread in the measured values and that, consequently, genes differentially expressed at a much lower level may indeed be statistically significant. On the other hand, for some genes where there is much greater variation in the measurement of the $\log_2(\text{ratio})$, two-fold may not represent a significant difference.

A more statistically defensible approach is to calculate the mean μ and standard deviation σ for the distribution of $\log_2(\text{ratio})$ values. Differential expression at the 95% confidence level can then be identified as $\log_2(\text{ratio})$ values more than 1.96 standard deviations from the mean.

3.5.1 Intensity-dependent estimation of differential expression

Even the approach described above may misidentify differentially expressed genes. While Lowess normalisation removes dye-specific artefacts that appear for low intensity data points, the dataset exhibits additional structure that can be used to evaluate patterns of differential expression. An examination of the R–I plot in Figure 3.5 suggests that $\log_2(\text{ratio})$ values at lower intensities are much more variable than those measured at higher intensities. In an R–I plot, a fixed $\log_2(\text{ratio})$ threshold is represented by a pair of straight, horizontal, parallel lines. Such a fixed cut-off does not take into account the observed intensity dependence.

An alternative is to identify differentially expressed genes using an intensity-dependent approach. Using a sliding window, one calculates the local mean and standard deviation of the $\log_2(\text{ratio})$ for each data point in the normalised R–I plot. This can then be used to calculate a Z-score, which simply measures the number of standard deviations each data point is from the mean. Differentially expressed genes at the 95% confidence level then have a value of $|Z| > 1.96$.

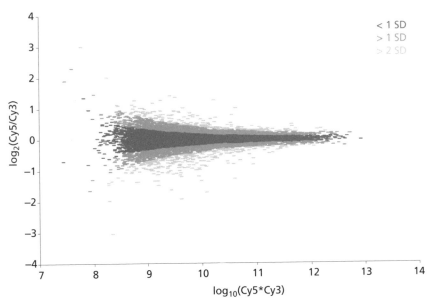

Figure 3.8 Intensity-dependent identification of differential expression. Data within and outside one- and two-standard deviation limits are shown in dark grey ($|Z| < 1$), medium grey ($1 < |Z| < 2$) and light grey ($|Z| > 2$), respectively. The variation in differential expression measurements depends on the intensity. Consequently, the fold-change level that is statistically significant at high intensities may not be significant at lower intensities. By computing an intensity-dependent Z-score, we can identify differentially expressed genes with high confidence, taking into account the systematic variation in the data. Many microarray experiments use an arbitrary, global fold-change level to identify differentially expressed genes; in an R–I plot, these would be parallel horizontal lines. However, as illustrated in the R–I plot (Figure 3.5), the significance that one can attach to differential expression depends on the intensity. (Yang *et al.*, 2002a)

Figure 3.8 depicts the results of such a calculation: data within and outside one- and two-standard deviation limits are shown in dark grey ($|Z| < 1$), medium grey ($1 < |Z| < 2$) and light grey ($|Z| > 2$), respectively; genes greater than two standard deviations from the local mean ($|Z| > 2$) are identified as being significantly differentially expressed at 95.5% confidence.

3.5.2 Analysis of variance

An alternative approach that has been advocated for the identification of differentially expressed genes is the *analysis of variance* (ANOVA). ANOVA is a powerful tool for the analysis of large datasets that can provide estimates of the relative effects contributing to an observation and error estimates for each of these. The purpose of ANOVA is to test for significant differences between means by comparing variances. More specifically, by partitioning the total variation into different sources (associated with the different effects in the design), we are able to compare the variance due to the between-groups (or treatments) variability with that due to the within-group (treatment) variability. Under the null hypothesis that there are no mean differences between groups or treatments in the population, the variance estimated from the within-group (treatment) variability should be about the same as the variance estimated from the between-groups (treatments) variability. Significantly differentially expressed genes are those that violate this null hypothesis.

In general, the purpose of ANOVA is to test for significant differences between means (the name is derived from the fact that in order to test for statistical significance between means, we are actually analysing variances). ANOVA is a generalisation of better-known statistical techniques, and if we are comparing only two means, ANOVA will give the same results as a *t*-test. At the heart of ANOVA is the assumption that variances in the measurements made in an experiment can be divided, or partitioned, into those arising from different sources. The variance, you may recall, is the sum of squared deviations from the overall mean divided by the sample size minus 1, or

$$s^2 = \frac{\sum_{i=1}^{N}(x_i - \mu)^2}{(N-1)}$$

where N is the sample size and μ is the sample mean. Thus, given a certain N, the variance is a function of the *sums of squares* (actually squared deviates), or SS. The goal of an ANOVA is to compare the SS measured within each group with the variability due to the differences between means.

In a microarray experiment, we want to evaluate the relative contributions of the genes, the arrays constructed from them, the dyes used to label the samples, and the RNA samples used in the assay. Kerr, Churchill and collaborators at the Jackson Laboratory (Kerr and Churchill, 2001; Kerr *et al.*, 2000; Yang *et al.*, 2002a) have developed an ANOVA model that provides an estimate of the various potential contributions from a variety of sources to the observed hybridisation intensities. Let y_{ijkg} be the fluorescence intensity measured from *Array* (the slide) i, using *Dye j*, to label RNA sample (called here the *Variety*, due to the fact that this approach was first used in the analysis of crop varieties grown on different plots) k, and using a feature to represent a *Gene g*, on the appropriate scale (such as the log of the measured intensity). The basic premise is that what one actually wants to measure – the expression of the mRNA representing a particular gene on the array – is in fact influenced by a variety of factors that contribute random and systematic noise to the measurements. One then constructs an ANOVA model representing how these primary terms and interactions contribute to the observed measurement, such as

$$y_{ijkg} = y_0 + A_i + D_j + (AD)_{ij} + G_g + (AG)_{ig} + (VG)_{kg} + (DG)_{jg} + \varepsilon_{ijkg}$$

Here y_0 is an array-independent constant, $(AD)_{ij}$ represents dye-specific effects limited to a single array, $(AG)_{ig}$ is a gene-specific effect on a particular array such as a printing irregularity on a single array, $(DG)_{jg}$ is an apparent dye–gene interaction such as a specific feature which always appears green in a two-colour hybridisation, ε_{ijkg} is an additional random error assumed to be normally distributed around 0, and $(VG)_{kg}$ is the effect you actually want to measure, the true measure of the expression of a particular gene in the labelled extract. In this model, y_0, A, D and AD are overall normalisation terms. Using well-established techniques, one can use this model to calculate the relative effects of the various contributors to the observed fluorescence intensities, and arrive at an estimate with errors of the gene-specific expression for each experiment. The result is a collection of $(VG)_{kg}$ measurements that represent the $\log_2$(ratio) values. Each of these estimates the expression for a gene in a particular sample relative to its average across all measurements. Differentially expressed genes can then be identified using an F-test, a generalisation of the better-known t-test: the F-test compares means across multiple samples to identify those genes exhibiting significant deviation in measured expression from the mean.

ANOVA is particularly effective in the analysis of experiments employing a loop design, where there is no natural reference sample that can serve as a basis for comparison between samples.

References

Chatterjee, S., and Price, B. (1991). *Regression Analysis by Example* (John Wiley & Sons, New York).

Chen, Y., Dougherty, E. R., and Bittner, M. L. (1997). Ratio-based decisions and the quantitative analysis of cDNA microarray images. *Journal of Biomedical Optics* 2, 364–374.

Cleveland, W. S. (1979). Robust locally weighted regression and smoothing scatterplots. *Journal of the American Statistical Association* 74, 829–836.

Hughes, T. R., Marton, M. J., Jones, A. R., Roberts, C. J., Stoughton, R., Armour, C. D., Bennett, H. A., Coffey, E., Dai, H., He, Y. D., *et al.* (2000a). Functional discovery via a compendium of expression profiles. *Cell* 102, 109–126.

Hughes, T. R., Roberts, C. J., Dai, H., Jones, A. R., Meyer, M. R., Slade, D., Burchard, J., Dow, S., Ward, T. R., Kidd, M. J., *et al.* (2000b). Widespread aneuploidy revealed by DNA microarray expression profiling. *Nature Genetics* 25, 333–337.

Irby, R., Chen, T., Chambers, A. F., Coppola, D., Quackenbush, J., Agrawal, D., and Yeatman, T. J. (2002). Expression profiling identifies oesteopontin as a new marker of colon cancer progression that augments tumorigenicity. *Journal of the National Cancer Institute* 94, 513–521.

Kerr, M. K., and Churchill, G. A. (2001). Statistical design and the analysis of gene expression microarray data. *Genetical Research* 77, 123–128.

Kerr, M. K., Martin, M., and Churchill, G. A. (2000). Analysis of variance for gene expression microarray data. *Journal of Computational Biology* 7, 819–837.

Lee, P. D., Sladek, R., Greenwood, C. M. T., and Hudson, T. J. (2002). Control genes and variability: absence of ubiquitous reference transcripts in diverse mammalian expression studies. *Genome Research* 12, 292–297.

Li, C., and Wong, W. H. (2001). Model-based analysis of oligonucleotide arrays: Expression index computation and outlier detection. *Proceedings of the National Academy of Sciences of the United States of America* 98, 31–36.

Pritchard, C. C., Hsu, L., Delrow, J., and Nelson, P. S. (2001). Project normal: defining normal variance in mouse gene expression. *Proceedings of the National Academy of Sciences of the United States of America* **98**, 13266–13271.

Schadt, E. E., Li, C., Ellis, B., and Wong, W. H. (2001). Feature extraction and normalization algorithms for high-density oligonucleotide gene expression array data. *Journal of Cell Biochemistry (Supplement)* **37**, 120–125.

Thellin, O., Zorzi, W., Lakaye, B., De Borman, B., Coumans, B., Hennen, G., Grisar, T., Igout, A., and Heinen, E. (1999). Housekeeping genes as internal standards: use and limits. *Journal of Biotechnology* **75**, 291–295.

Tseng, G. C., Oh, M.-K., Rohlin, L., Liao, J. C., and Wong, W. H. (2001). Issues in cDNA microarray analysis: quality filtering, channel normalization, models of variations and assessment of gene effects. *Nucleic Acids Research* **29**, 2549–2557.

Yang, I. V., Chen, E., Hasseman, J. P., Liang, W., Wang, S., Sharov, V., Saeed, A. I., White, J., Li, J., Lee, N. H., *et al.* (2002a). Within the fold: Assessing differential expression measures and reproducibility in microarray assays. *Genome Biology* **3**(11): RESEARCH 0062.

Yang, Y. H., Dudoit, S., Luu, P., Lin, D. M., Peng, V., Ngai, J., and Speed, T. P. (2002b). Normalization for cDNA microarray data: a robust composite method addressing single and multiple slide systematic variation. *Nucleic Acids Research* **30**, e15.

Yang, Y. H., Dudoit, S., Luu, P., and Speed, T. P. (2001). Normalization for cDNA microarray data. *Proceedings of SPIE, BIOS 2001, Microarrays: Optical Technologies and Informatics* **4266**, 141–152.

CHAPTER 4

Analysis of gene expression data matrices

Science is built with facts as a house is with stones – but a collection of facts is no more a science than a heap of stones is a house.

Jules H. Poincaré

4.1 Introduction

The goal of microarray data analysis is to find relationships and patterns in the data and ultimately achieve new insights into the underlying biology. For instance, one could look for groups of genes having similar expression under similar conditions and try to find whether their products share similar functional roles in the cell, or for genes whose expression depends on the particular state of the system and see if the functions of their products can help to explain the particular phenotype.

As discussed earlier, it is convenient to split microarray data analysis into two stages – transforming the raw data into a gene expression matrix, and analysis of the gene expression matrix (Table 4.1). The first stage was discussed in the previous chapter. In this chapter our starting point is the gene expression data matrix, where rows represent genes, columns represent *experimental conditions* or *samples*, and the values at each position in the matrix characterise the expression level of the particular gene under the particular experimental condition. We will

Table 4.1 An overview of some of the steps typically carried out in the analysis of microarray data. Data analysis can be divided into two stages – transformation of the raw data into a gene expression matrix (which was covered in Chapter 3) and analysis of the gene expression matrix (which is the subject of the present chapter). The specific methods used at each step, or whether the step is omitted, depends on the nature of the samples, the microarray platform and the objectives of the analysis.

Stage	Data description	Objective	Example of data transformation
Transformation of the raw data into a gene expression matrix	Primary data capture	Obtain raw gene expression measurements	
	Image processing	Obtain quantifiable description of signal intensity associated with each spot	Align grid, calculate sum of pixel intensities for each spot
	Corrected signal intensity	Correct for background effects and variability between spots	Subtract background for each spot, local normalisation
	Normalisation	Select 'high quality' data/flag 'low confidence' data	Filter, set floors/ceilings, threshold values
		Transform data to account for non-biological variability between samples	Global normalisation
Analysis of the gene expression matrix	Processed data	Transform data to enhance characteristic(s) of interest	Calculate fold difference/change in gene expression, log transformation, filter, mean centre, adjust variance to 1, identify differentially expressed gene
		Group data based on characteristics of interest	Clustering, principal component analysis
		Relate subsets of grouped data to information from other sources	Gene and sample annotation

call these values *gene expression levels*. (Here we will use the terms 'a sample' and 'an experimental condition' as synonyms denoting columns in the matrix, although in reality an experimental condition can be represented in the matrix by several replicate samples, i.e. by several columns.) We will refer to rows of values in the matrix as *gene expression profiles*, and to the columns as *sample expression profiles*.

To achieve our ultimate goal of obtaining new insights into biology, at some point we will need to use additional biological information about the genes and the experimental conditions which is not directly represented in the gene expression matrix. This information can, however, be added to the matrix in the form of gene and sample annotation. For instance, gene annotation may include the gene names and sequence information, location in the genome, a description of the functional roles for known genes, and links to the respective entries in gene sequence databases. Sample annotation may provide information about the part of the organism from which the sample was taken or which cell type was used, or whether the sample was treated, and if so what was the treatment (e.g. a chemical compound and concentration). Samples may also be related: for instance, they may form a time course. The gene expression matrix together with the annotation will be called the *annotated gene expression data matrix*. We can regard gene annotations as additional columns and sample annotations as additional rows (Figure 1.3).

Depending on how we treat the annotation, gene expression data analysis can be either *supervised* or *unsupervised*. In supervised analysis we use the annotation from the very beginning. A typical example of supervised analysis is sample classification. Here we use sample annotation to split the set of samples into two or more classes, for instance 'healthy' or 'diseased' tissues, and try to find patterns or features in the expression data that are characteristic of each of the individual classes. If such features are found, they can be used for diagnostics, i.e. to attribute a 'healthy' or 'diseased' label to new samples. This is illustrated on the left side of Figure 4.1, where the objects are dots annotated as filled or hollow shown in two-dimensional space. The task of the classification is to find a way to separate them. A straight line can separate the filled and hollow dots with only few classification mistakes. However, the quality of a classifier is determined by how well it performs when new objects are introduced.

Unsupervised analysis is based on looking for structure in the data itself, ignoring any annotation. Examples of such analysis are gene clustering (finding sets of genes with similar expression patterns), sample clustering (finding which samples are similar in terms of similarly expressed genes) and principal component analysis (finding the axes of greatest variability). Annotation information is taken into account only later, for instance to see whether the clusters of similarly expressed genes contain those with similar functional roles. This is illustrated on the right side of Figure 4.1, where we can identify clusters of dots that are closer to each other than to those of other clusters.

A combination of supervised and unsupervised analysis is possible – information found in annotations can be used only to guide the analysis: for instance,

(a)   (b)

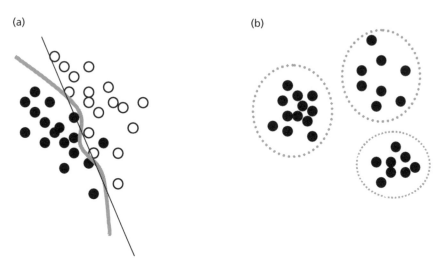

Figure 4.1 Supervised vs. unsupervised analysis. The task of supervised analysis (a) is to find a 'classifier' separating known classes (filled and hollow dots). The task of unsupervised analysis (b) is to find groups inherent to the data, such as clusters of points that are closer to each other than to the rest.

labels 'healthy' or 'diseased' can be added to the matrix columns and used to guide the clustering. This is often called partially supervised analysis.

Specific properties of the gene expression matrix can be taken into account in the analysis. For instance, if some of the experimental conditions represented in the matrix are closely related (e.g. replicates), the gene expression values in the columns are likely to be similar. Depending on the goal of the particular analysis, it is possible that the contributions from such columns should be downweighted. Another example is the analysis of time course data, when we can look for possible periodicity in gene expression patterns. It would be meaningless to look for periodicity if the list of columns were, for instance, chemical treatments.

Finally, we can distinguish between hypothesis driven analysis and exploratory data analysis (the latter is sometimes known as data mining). Gene clustering by expression profile similarity can be considered an example of exploratory analysis – before the analysis is performed we may not know if there are clusters in the data, how many clusters, whether there is any hierarchical structure in the clustering, and whether the possible clusters have any correlation with other biological information. An example of the hypothesis driven approach is taking a gene of particular interest, and finding its expression pattern and which other genes are expressed similarly. Note that the distinction between exploratory analysis and a hypothesis driven approach is not clear-cut.

Instead of giving a 'final answer' to a biological question, computational analysis methods generally only produce a hypothesis or can help to narrow down an existing hypothesis. Therefore the computational analysis (information processing) is only one step in an iterative loop of knowledge acquisition. To confirm or reject these hypotheses new experiments (material processing) are usually needed.

In this chapter we will discuss some of the common methods for gene expression data matrix analysis. The chapter should not be regarded as a comprehensive list of all the available analysis methods, nor as a literature review – its main goal is to provide an understanding of the basic principles underlying these methods. More details about many of the described methods can be found in various multivariate analysis textbooks, for instance Everitt and Dunn (2001).

4.2 Gene expression data matrices: their features and representations

4.2.1 Gene expression matrices

As we already mentioned, the starting point in this chapter is a gene expression data matrix, which can be either annotated or unannotated. Effectively we are trying to separate as far as possible the gene expression data matrix analysis from the information processing leading from the raw images to the matrix. For instance, we are no longer interested in which image analysis algorithm has been used to analyse the raw data, and even what normalisation methods have been used.

Although such abstraction is central to the concept of the gene expression data matrix, unfortunately we cannot ignore the nature of the gene expression values completely. For instance, we may need to know: do the columns represent absolute or relative measurements, are the values in the matrix ratios, log ratios or something else, are the reference samples for relative measurements the same for all the columns, and has the normalisation been performed in a way that permits comparisons between different columns of the matrix? Depending on the answers to these questions various analysis approaches may or may not make sense.

We have to distinguish between matrices that contain data from a set of related hybridisations that were done using the same protocol, the same reference sample, and the same batch of arrays, from matrices constructed by combining results from different experiments using different reference samples. Often gene expression matrices are a combination of both, i.e. they contain different sections where (inside a section) the same reference sample has been used. For instance, the well-known and frequently studied dataset of 80 experimental conditions in yeast (Eisen *et al.*, 1998) was composed by putting together various independent time series, including diauxic shift (the stage at which yeast switch from anaerobic to aerobic growth) and cell cycle data. We will discuss this question in more detail later in the chapter (see Section 4.3.11). A special type of gene expression matrix is produced from the circular, or 'loop', design (see Section 3.4.4).

Consider the gene expression matrices in Tables 4.2, 4.3 and 4.4. They all represent the expression levels of genes G1–G8 for experimental conditions C1, C2, C3 and C4, but use different measurements – absolute levels (in abstract units) in Table 4.2, relative measurements using C4 as a reference in Table 4.3, and relative values transformed into logarithms in Table 4.4. Note that gene expression profiles that look rather different in the absolute value table may be similar in the relative measurement table.

Most gene expression data analysis algorithms assume that the gene expression values are represented as *scalars*, i.e. one numerical value (such as log ratio) per expression value. If the gene expression levels for a specific experimental condition have been measured, for instance, in three replicate hybridisations, then to use such an algorithm either they have to be treated as three separate experimental conditions (i.e. all three columns should be treated separately) or they have to be replaced by one generalising scalar such as arithmetic mean or median. In the latter case all the information about the variance is lost, and so it may be advisable to include all the replicate measurements in the matrix as separate columns. However, it should be noted that columns for replicate experiments are likely to be closely correlated, and therefore if some of the samples are represented in data by more replicates than others, they will contribute to gene expression profiles more and may skew the clustering.

For relative measurements, where we typically rely on ratios or log ratios, any information about the absolute level of gene expression is lost (e.g. ratio values for 400/200, 40/20 and 4/2 are all the same). Knowing the respective absolute intensity values can provide the algorithm with valuable information about the reliability of the ratio. Yet most current data analysis algorithms (e.g.

Table 4.2 Gene expression data matrix of absolute expression measurements for samples C1, C2, C3 and C4, given in abstract measurement units.

	C1	C2	C3	C4
G1	2	3	4	2
G2	2	3	4	20
G3	20	30	40	20
G4	200	300	400	200
G5	20	30	40	2
G6	20	30	40	1
G7	1	4	16	4
G8	4	5	4	4

Table 4.3 Gene expression data matrix containing relative expression measurements comparing conditions C1, C2 and C3 with the control condition C4. Compare with Table 4.2. Note that rather different absolute measurements, such as 4/2, 40/20 or 400/200, all give the same ratios.

	r1 = C1/C4	r2 = C2/C4	r3 = C3/C4
G1	1	1.5	2
G2	0.1	0.15	0.2
G3	1	1.5	2
G4	1	1.5	2
G5	10	15	20
G6	20	30	40
G7	0.25	1	4
G8	1	1.25	1

Table 4.4 Gene expression data matrix containing log ratios (base 2) from Table 4.3. Compare with Table 4.3. Note that four-fold down- and up-regulation for gene G7 reflects as 0.25 and 4 in the ratio table, while in the log ratio table the values are symmetric, −2 and +2.

	log r1	log r2	log r3
G1	0	0.58	1
G2	−3.32	−2.74	−2.32
G3	0	0.58	1
G4	0	0.58	1
G5	3.32	3.91	4.32
G6	4.32	4.91	5.32
G7	−2	0	2
G8	0	0.32	0

clustering algorithms) ignore this information. Currently the use of this information is difficult, because of the lack of established error models and standards for the measurement and representation of gene expression levels.

The use of ratios (or log ratios) instead of absolute gene expression values scales the values in the gene expression matrix towards a more comparable range. For instance, the absolute gene expression levels (e.g. measured as the number of mRNA copies per cell) have a wide range, from less than one copy per cell on average to hundreds of mRNA copies per cell. In ratio matrices these values will usually be compensated if the genes that tend to have high expression levels are likely to be present in higher amounts in the reference sample as well.

As a starting point of gene expression data analysis we can also use a discretised gene expression matrix, such as a binary matrix where 0 means that the gene is not expressed, and 1 that it is expressed. For comparative experiments we can use three values: −1 meaning substantially reduced expression compared with the reference sample, 0 unchanged expression and +1 substantially increased expression (Table 4.5). Although such discretisation may result in the loss of information from the original table, it enables the application of various analysis methods that are not appropriate for real value matrices. For instance, mutual information between genes or conditions (see Section 4.2.3.4) can be defined for discretised tables more easily than for real value matrices.

Finally, it should be noted that in gene expression data obtained from microarray experiments there may be missing (unknown or formally *undefined*) values. Some of the analysis methods cannot handle matrices with undefined

Table 4.5 Discretised gene expression matrix obtained from the log ratios in Table 4.4 by treating all values of at least 2 as up-regulation (represented by 1), those of less than or equal to −2 as down-regulation (represented by −1), and others as unchanged (represented by 0). D(r) denotes the above described discretisation of the log ratio.

	D(r1)	D(r2)	D(r3)
G1	0	0	0
G2	−1	−1	−1
G3	0	0	0
G4	0	0	0
G5	1	1	1
G6	1	1	1
G7	−1	0	1
G8	0	0	0

values, and to use these methods either we have to exclude any row or columns containing an undefined value (which often leaves one with a rather small matrix) or find a way to interpolate the known values to fill in the missing ones.

Having organised the data in a gene expression matrix, a typical next step is to try to narrow down the genes of interest before more sophisticated analysis is attempted. The genes that are typically used are those that undergo a specified fold change in at least one of the samples, e.g. genes whose expression changes by at least two-fold.

4.2.2 Representation of expression data as vector space – sample space and gene space

Each gene (each row in the matrix) can be considered as a point in m-dimensional space, where m is the number of samples (columns in the matrix). Similarly, each sample (each column in the matrix) can be considered as a vector in n-dimensional space, where n is the number of genes (i.e. rows in the matrix). In this way we can talk about *gene space* and *sample* (or *condition*) *space*.

Let us consider an example of three genes A, B and C and two experimental conditions C1 and C2 (i.e. $m = 2$ and $n = 3$) (Table 4.6). This can be visualised either as 3 two-dimensional vectors in the condition space, or 2 three-dimensional vectors in the gene space, as depicted in Figure 4.2.

Each point in a multidimensional space defines a *vector* (joining the points in the respective space to the coordinate zero point). Viewing genes or samples as vectors or points in the appropriate multidimensional space allows one to use data analysis methods developed in linear algebra and to visualise different data transformations as operations in the respective vector space. Many of the properties of multidimensional space can be demonstrated in two or three dimensions, which is a particular attraction of this approach. For instance, we can visualise the similarity or difference between two gene expression profiles as the distance between the respective points or vectors.

When introducing new concepts related to multidimensional space, wherever possible we will demonstrate them in two or three-dimensional space first. The aim of this section is not to introduce a list of formulae, but to help the

Table 4.6 Gene expression matrix of three genes under two conditions. The gene expression measurements are in arbitrary units.

	C1	C2
A	2	3
B	3	4
C	4	2

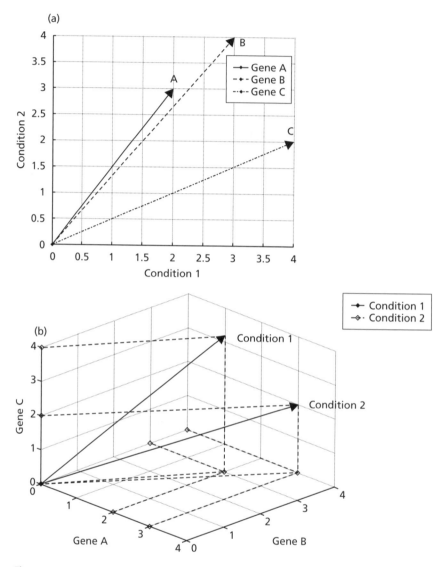

Figure 4.2 Visualising genes in condition space (a) and conditions in gene space (b) for the gene expression matrix given in Table 4.6.

reader to understand the meaning of the basic concepts used in expression profile comparison and how they relate to each other.

Throughout the rest of this chapter we will use the following notation. Let $\mathbf{X}$ be the gene expression matrix with m columns and n rows (we often call such a matrix an $m \times n$ matrix). Let x_{ij} be the expression value in the ith row and the jth column, i.e.

$$\mathbf{X} = \begin{pmatrix} x_{11} & x_{12} & \cdots & x_{1m} \\ x_{21} & x_{22} & \cdots & x_{2m} \\ \cdots & \cdots & \cdots & \cdots \\ x_{n1} & x_{n2} & \cdots & x_{nm} \end{pmatrix}$$

For instance, the expression of the three genes under two conditions in Figure 4.2 is represented by the 2×3 matrix

$$\begin{pmatrix} 2 & 3 \\ 3 & 4 \\ 4 & 2 \end{pmatrix}$$

The rows or columns of the matrix define vectors $\mathbf{A} = (a_1, \ldots, a_k)$ (e.g. $\mathbf{A}_i = (x_{i1}, \ldots, x_{im})$ for the ith row of the matrix and $\mathbf{A}_j = (x_{1j}, \ldots, x_{nj})$ for the jth column). Given a vector $\mathbf{A} = (a_1, \ldots, a_k)$, we define its length $|\mathbf{A}|$ as

$$|\mathbf{A}| = \sqrt{a_1^2 + \ldots + a_k^2}$$

For instance, the length of the vector $(2, 3, 4)$ is $\sqrt{29}$. Note that in the case of two-dimensional space, i.e. for $k = 2$, the formula expresses the length of the hypotenuse in accordance with the well-known Pythagoras's theorem.

If we are given a vector $\mathbf{A} = (a_1, \ldots, a_k)$ of length $|\mathbf{A}|$, we can transform it to vector $\mathbf{A}' = (a_1', \ldots, a_k')$ having the same direction but unit length, by taking $a_1' = a_1/|\mathbf{A}|, \ldots, a_k' = a_1/|\mathbf{A}|$ (Figure 4.3). This is sometimes called *vector normalisation* and should not be confused with the normalisation of micro-array data, discussed in the previous section.

4.2.3 Distance and similarity measures in expression space

Most of the gene expression data analysis methods are based on comparisons between the gene or sample expression profiles. In order to make these comparisons first we need a way to measure similarity or dissimilarity between these objects, i.e. between vectors representing genes or samples. Often it is easier to measure the distance between the objects (vectors in our case) instead of the similarity, though one can be transformed into the other.

The distance between A and B, $D(A, B)$, is said to be *metric* if it satisfies the following properties:

1 if $A = B$, then $D(A, B) = 0$, i.e. the distance from an object to itself is 0;
2 if $A \neq B$, then $D(A, B) \geq 0$, i.e. the distance is always non-negative;
3 $D(A, B) = D(B, A)$, i.e. it does not matter in which order we measure the distance;
4 $D(A, B) + D(B, C) \geq D(A, C)$, i.e. given three objects, the length of a direct path from the first to the third objects cannot be greater than the length of the path through the second object.

The last is called the *triangle inequality*. The distance measures that satisfy properties (1)–(3), but not the triangle property, are called *semimetric*. Some data analysis algorithms can be applied only for metric distances. An example of non-metric distance measure is a distance measured between cities as the cheapest airfare to fly from one city to another.

4.2.3.1 EUCLIDEAN, MINKOWSKI, MANHATTAN, ANGLE AND CHORD DISTANCES

Euclidean distance is the most common distance measure, and the one we use in everyday situations, e.g. to measure how long a cable is needed to reach the corners of a room from a socket. Euclidean distance between points $A = (a_1, a_2)$ and $B = (b_1, b_2)$ in two dimensions can be expressed using Pythagoras's theorem:

$$D_{\text{Eucl}}(\mathbf{A}, \mathbf{B}) = \sqrt{(a_1 - b_1)^2 + (a_2 - b_2)^2}$$

For instance, for genes $A = (2, 3)$ and $B = (3, 4)$ in Figure 4.2 the distance between them equals $\sqrt{(2-3)^2 + (3-4)^2} = \sqrt{2}$. In n-dimensional space for vectors $\mathbf{A} = (a_i, \ldots, a_n)$ and $\mathbf{B} = (b_i, \ldots, b_n)$, Euclidean distance can be expressed as

$$D_{\text{Eucl}}(\mathbf{A}, \mathbf{B}) = \sqrt{\sum_{i=1}^{n}(a_i - b_i)^2}$$

i.e. the square root of the sum of the square of the distances between the points in each dimension. *Minkowski* distance is a generalisation of Euclidean distance and is expressed as

$$D_{\text{Minkowski}}(\mathbf{A}, \mathbf{B}) = \left(\sum_{i=1}^{n} |a_i - b_i|^p \right)^{1/p}$$

The parameter p is called the *order*: the higher the value of p, the more significant is the contribution of the largest components $|a_i - b_i|$. Euclidean distance corresponds to the distance described by the second order Minkowski distance (i.e. $p = 2$).

Manhattan, or rectilinear, distance corresponds to the distance described by the first order Minkowski distance, when $p = 1$. The Manhattan distance between the coordinates of the two vectors $\mathbf{A} = (a_1, \ldots, a_n)$ and $\mathbf{B} = (b_1, \ldots, b_n)$ is the sum of the distance between them in each dimension

$$D_{\text{Manhattan}}(\mathbf{A}, \mathbf{B}) = \sum_{i=1}^{n} |a_i - b_i|$$

In two-dimensional space Manhattan distance is the distance between the points on the first axis, plus the distance between them on the second axis (e.g.

the Manhattan distance between genes A and C in Figure 4.2 equals $|4 - 2| + |2 - 3| = 3$). Manhattan distance is sometimes referred to as 'city block distance' as it measures the route one might have to travel between two points in a place such as Manhattan where the streets and avenues are arranged at right angles to one another. It is also known as *Hamming distance* when applied to data expressed in binary form, e.g. if the expression levels of the genes have been discretised into 1s and 0s.

Euclidean distance is one of the most intuitive ways to measure the distance between points in space, but it is not always the most appropriate one for expression profiles. For example, if the gene expression data represent comparative expression measurements, the absolute values in the matrix may not be meaningful. Even with absolute measurements sometimes we may be more interested in comparing the changes in the expression, rather than the absolute levels. In these cases we need to define distance measures that score as similar gene expression profiles that show similar trends, rather than those that depend on the absolute levels.

Two simple distance measures that can be used in such cases are the angle and chord distances. Let us first assume that all the expression values are positive. The *angle distance* is defined as the angle (e.g. expressed in radians: 360° equals 2π radians) between the two vectors. For positive values the angle varies in the range from 0 to $\pi/2$. Instead of the angle α, we can also use function $\sin \alpha$, which for positive vectors varies in the range from 0 to 1. Note that $\sin^2 \alpha = 1 - \cos^2 \alpha$.

Let us consider Figure 4.3. The cosines of the angle between the two vectors **A** and **B** can be expressed as $\cos \alpha = \cos(\gamma - \beta) = \cos \beta \cos \gamma + \sin \beta \sin \gamma$. Noting that $\cos \beta = a_1/|\mathbf{A}|$, $\sin \beta = a_2/|\mathbf{A}|$, $\cos \gamma = b_1/|\mathbf{B}|$, and $\sin \gamma = b_2/|\mathbf{B}|$ we obtain

$$\cos \alpha = \frac{a_1 b_1 + a_2 b_2}{|\mathbf{A}||\mathbf{B}|}$$

The angle between the normalised vectors $\mathbf{A}' = (a_1', a_2')$ and $\mathbf{B}' = (b_1', b_2')$ is the same as the angle between the original vectors, therefore

$$\cos \alpha = a_1' b_1' + a_2' b_2'$$

In n-dimensional space for vectors $\mathbf{A} = (a_i, \ldots, a_n)$ and $\mathbf{B} = (b_i, \ldots, b_n)$ the cosine is defined as

$$\cos \alpha = \sum_{i=1}^{n} a_i' b_i' = \frac{\sum_{i=1}^{n} a_i b_i}{|\mathbf{A}||\mathbf{B}|}$$

The sum $\sum_{i=1}^{n} a_i b_i$ is known as the *dot product*, sometimes denoted by $\mathbf{A} \cdot \mathbf{B}$, and can be used as a similarity measure.

The *chord distance* is defined as the length of the chord between the vectors of unit length having the same directions as the original ones, as shown in

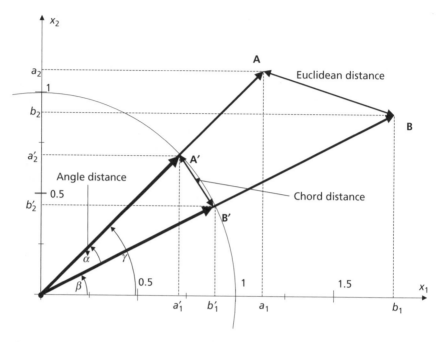

Figure 4.3 Angle and chord distances (see text for explanation).

Figure 4.3. Evidently the chord distance is the same for the normalised as for the original vectors. For normalised vectors the chord and the Euclidean distances are the same. This is an important property, since some of the data analysis methods, such as K-means clustering, require the use of Euclidean distance properties. If one first normalises the vectors, one can perform analysis in normalised space using Euclidean distance, which will give the same results as using chord distance in the original space.

Using Pythagoras's theorem, in the case of two dimensions, we see from Figure 4.3 that

$$D_{\text{chord}}(\mathbf{A}, \mathbf{B}) = \sqrt{(a_1' - b_1')^2 + (a_2' - b_2')^2} = \sqrt{\left(\frac{a_1}{|\mathbf{A}|} - \frac{b_1}{|\mathbf{B}|}\right)^2 + \left(\frac{a_2}{|\mathbf{A}|} - \frac{b_2}{|\mathbf{B}|}\right)^2}$$

Since $(a - b)^2 = a^2 - 2ab + b^2$, after simple transformations we obtain

$$D_{\text{chord}}(\mathbf{A}, \mathbf{B}) = \sqrt{2(1 - (a_1' b_1' + a_2' b_2'))} = \sqrt{2(1 - \cos\alpha)}$$

The last part of the equation is true for any number of dimensions. We can see directly from Figure 4.3 that $D_{\text{chord}}(\mathbf{A}, \mathbf{B}) = 2\sin\alpha/2$, confirming the well-known trigonometric relationship $2\sin^2\alpha/2 = 1 - \cos\alpha$.

A related distance measure that is sometimes used for comparing gene expression is the *chi-square metric* (χ^2 metric), which is used in correspondence

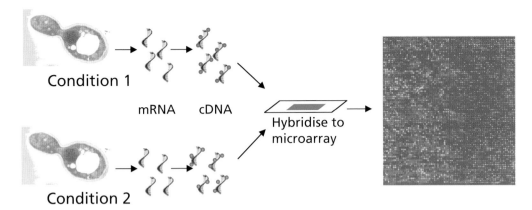

Plate 1.1 A typical microarray experiment. The mRNA is extracted from each of the populations of cells and a representation of the mRNA is labelled with one of two different labels, e.g. a green fluorescent dye for cells grown under condition 1 and a red fluorescent dye for cells grown under condition 2 (more precisely, the labelling is typically carried out by synthesising cDNA complementary to the extracted mRNA by reverse transcription). Both extracts are incubated with the microarray. Labelled gene products from the extracts preferentially bind to their complementary sequences; non-bound sample is removed by washing and the hybridised signal is detected by scanning.

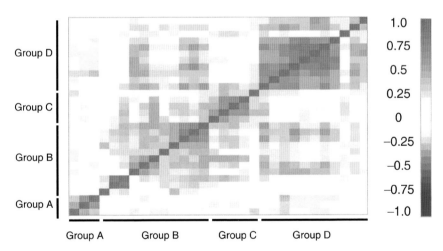

Plate 4.1 Pearson correlation. Here each of the small squares represents one dataset, comprising information about the expression levels of approximately 6000 genes in yeast. Each of the squares is coloured according to the similarity of the datasets compared. A Pearson correlation of 1.0 is represented as a red square and a perfect negative correlation is represented by a blue square, with 'no correlation' represented in white. Some of the datasets are more related to each other than others, as illustrated by the larger pink areas and the groupings of the datasets (the comparisons are carried out for every gene). For example, the datasets in group A are more related to each other than they are to the other datasets. Note that each dataset compared with itself has a perfect positive correlation, as shown by the red diagonal line.

facing p. 84

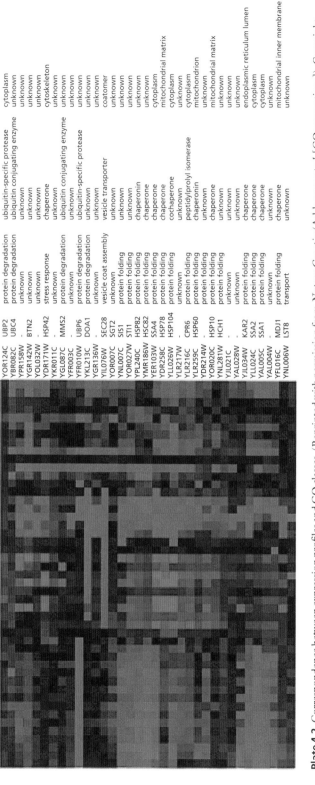

Plate 4.2 Correspondence between expression profiles and GO classes. (Reprinted with permission from *Nature Genetics* (Ashburner *et al.* [GO consortium]), Copyright 2000 Nature Publishing Group.)

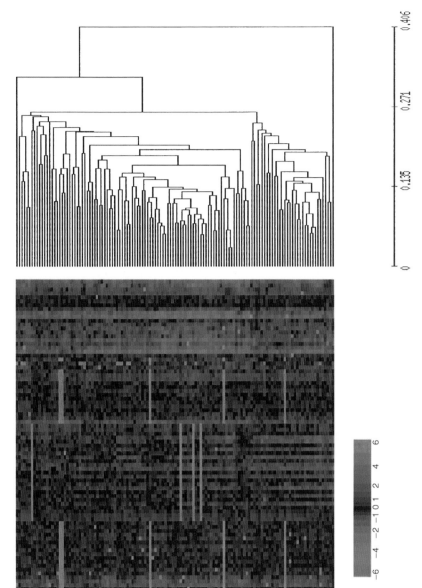

Plate 4.3 Dendrograms obtained from hierarchical clustering. The coloured representation of the data to the left of the dendrograms is known as a *heat map*. The colours simply represent the values in the gene expression matrix – positive values are encoded as red and negative ones as green. One can see that similar gene expression profiles (i.e. strings of similar colours) tend to be close together.

0' 60' 120' 240' 360'

Fold repression | Fold induction

>9 >3 1 >3 >9

Plate 4.4 Heat map of a hierarchical clustering. The change in gene expression over a time course is shown here as a heat map, as obtained using the visualisation tool Treeview. Each column corresponds to a different time point in the time course, with each row representing a single gene. The colours indicate the change in gene expression for each gene for that time point, with red indicating an increase in gene expression, black indicating no change, and green indicating a decrease in gene expression.

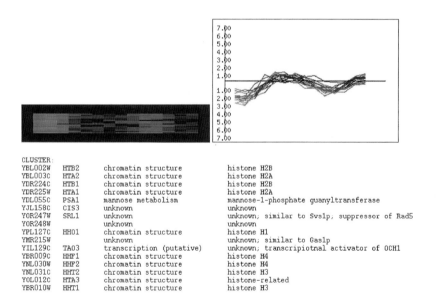

```
CLUSTER:
YBL002W   HTB2   chromatin structure       histone H2B
YBL003C   HTA2   chromatin structure       histone H2A
YDR224C   HTB1   chromatin structure       histone H2B
YDR225W   HTA1   chromatin structure       histone H2A
YDL055C   PSA1   mannose metabolism        mannose-1-phosphate guanyltransferase
YJL158C   CIS3   unknown                   unknown
YOR247W   SRL1   unknown                   unknown; similar to Svs1p; suppressor of Rad5
YOR248W          unknown                   unknown
YPL127C   HHO1   chromatin structure       histone H1
YMR215W          unknown                   unknown; similar to Gas1p
YIL129C   TAO3   transcription (putative)  unknown; transcripiotnal activator of OCH1
YBR009C   HHF1   chromatin structure       histone H4
YNL030W   HHF2   chromatin structure       histone H4
YNL031C   HHT2   chromatin structure       histone H3
YOL012C   HTA3   chromatin structure       histone-related
YBR010W   HHT1   chromatin structure       histone H3
```

Plate 4.5 Clusters represented by a heat map (left) and profile graphs (right). Note that profile graphs intuitively suggest the interpretation of data as a time course, which is not always correct.

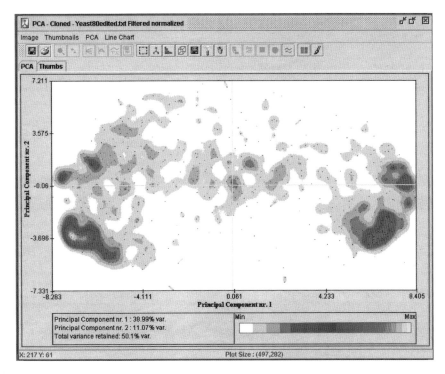

Plate 4.6 An example of a density plot in reduced space. (Figure generated using J-Express software).

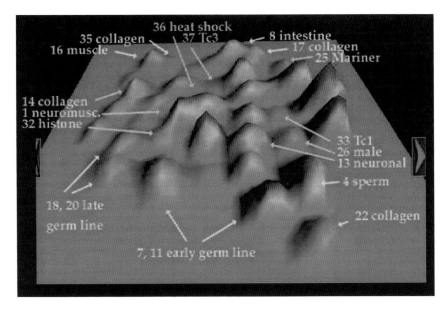

Plate 4.7 A gene expression terrain map created by VxInsight. The map shows 44 'gene mountains' representing genes whose expression is correlated out of 17,661 genes measured in 553 hybridisations. The altitude of a mountain corresponds to the density of the genes. (Reprinted with permission from *Science* (Kim *et al.*), Copyright 2001 American Association for the Advancement of Science.)

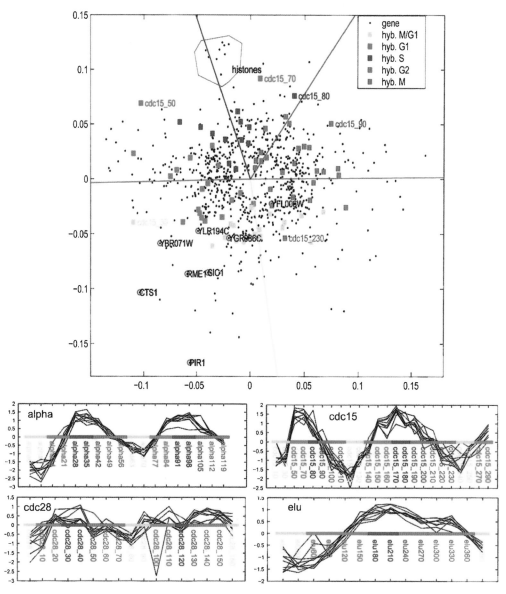

Plate 4.8 Correspondence analysis (see text for explanation). (Reprinted from *Proceedings of the National Academy of Sciences (USA)* (Fellenberg *et al.*), Copyright 2001 National Academy of Sciences, USA.)

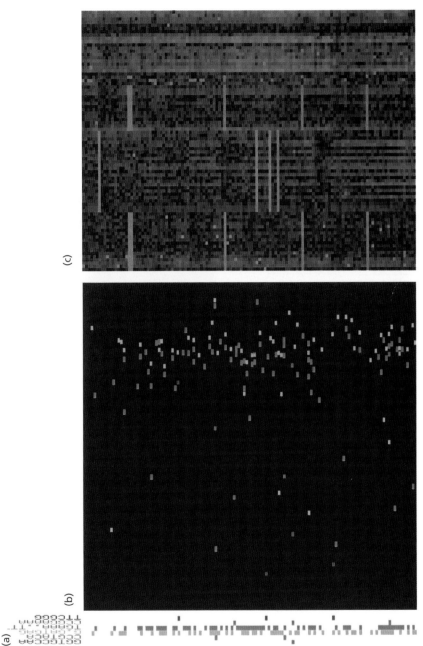

(a)

(b)

(c)

Plate 4.9 The figure shows DNA sequence and gene expression data for yeast genes where each row represents a particular gene. The coloured panel on the left shows (colour coded) DNA sequence elements, e.g. CGTGCCAA, found upstream of particular genes. The middle, predominantly black, panel shows the relative positions of these elements in relation to the 'start site', or ATG. The panel on the right represents the gene expression profile on the form of a heat map. (Figure generated using Expression Profiler Software.)

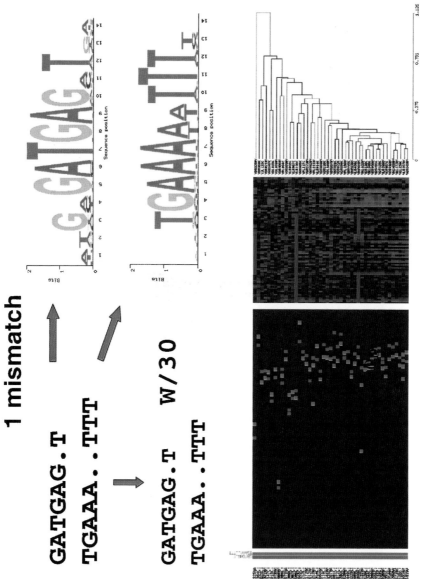

Plate 4.10 The occurrences of the sequence patterns GATGAG.T and TGAAA..TTT in the yeast genome upstream of genes in a tightly co-expressed cluster. Matching of statistically over-represented sequences was used to generate the 'pattern occurrence logos' on the top right of the figure. (One mismatch was permitted in carrying out the matching.)

analysis (described in Section 4.6). Consider the $m \times n$ matrix $\mathbf{X}$ in Section 4.2.2. The chord distance between rows $\mathbf{A} = (x_{a1}, \ldots, x_{am})$ and $\mathbf{B} = (x_{b1}, \ldots, x_{bm})$ equals

$$D_{chord}(\mathbf{A}, \mathbf{B}) = \sqrt{\sum_{j=1}^{m} \left(\frac{x_{aj}}{|\mathbf{A}|} - \frac{x_{bj}}{|\mathbf{B}|} \right)^2}$$

Suppose the sum of the elements x_{ij} of the column j equals $y_j = x_{1j} + \ldots x_{nj}$. In chord distance measure the samples (columns) that have the highest values y_i contribute predominantly (note that this is not always necessarily so, since differences between large values may be small). To compensate for this we can lessen the influence of these samples by 'normalising' the contributions from each experiment, i.e. by dividing each by the total expression level of the sample $y_j = x_{1j} + \ldots x_{nj}$:

$$D_{x^2}(\mathbf{A}, \mathbf{B}) = \sqrt{\sum_{j=1}^{m} \frac{1}{y_j} \left(\frac{x_{aj}}{|\mathbf{A}|} - \frac{x_{bj}}{|\mathbf{B}|} \right)^2}$$

Note that for some datasets this may lead to 'overcompensation', i.e. it is possible that for large sums y_j, the actual differences in parentheses are in fact small.

One can also adjust weights of various columns (i.e. contributions from various dimensions) by using *a priori* knowledge, e.g. by increasing the weights for those experimental conditions that are thought to be more important for determining the similarity between genes and decreasing the weights for the less important ones.

Finally, note that if we have defined a distance measure and are given gene expression matrix $m \times n$ (m samples and n genes), we can compute distances among all given expression profiles (vectors of length m), and represent these in a square $m \times m$ matrix $\mathbf{Z}$, with elements z_{ij} representing the distance between gene i and gene j. Since according to the definition of a distance $z_{ij} = z_{ji}$, the matrix is symmetric, and, since $z_{ii} = 0$, the diagonal elements in this matrix are zero.

4.2.3.2 PEARSON CORRELATION DISTANCE, ADJUSTING THE MEAN AND VARIANCE, CORRELATION MATRICES, AND THE RELATIONSHIP BETWEEN EUCLIDEAN AND CORRELATION DISTANCES

Although so far we have considered vectors with positive values, our considerations are generally true also for negative values. We obtain positive values for absolute expression measurements, but not for log ratio matrices. In the case where data represent log ratio matrices, it is sometimes natural to assume that the 'centre of gravity' for each gene expression profile is zero, which is equivalent to assuming that the total change in expression over all the experiments averages to zero. More precisely, we assume that the arithmetic mean of each gene expression profile is zero. We will see that under this assumption the angle distance is closely related to the Pearson correlation coefficient (see below).

First let us consider the example shown in Figure 4.4, where two expression profiles A and B for four samples are given. These are represented by vectors in four-dimensional space: $\mathbf{A} = (a_1, a_2, a_3, a_4)$ and $\mathbf{B} = (b_1, b_2, b_3, b_4)$. We can calculate the mean value for each profile as

$$\bar{a} = (a_1 + a_2 + a_3 + a_4)/4 \quad \text{and} \quad \bar{b} = (b_1 + b_2 + b_3 + b_4)/4$$

and shift each profile 'down' by its mean, i.e. obtain new vectors

$$\mathbf{A}^0 = (a_1 - \bar{a}, a_2 - \bar{a}, a_3 - \bar{a}, a_4 - \bar{a}) \quad \text{and} \quad \mathbf{B}^0 = (b_1 - \bar{b}, b_2 - \bar{b}, b_3 - \bar{b}, b_4 - \bar{b})$$

Their dot product equals

$$\mathbf{A}^0 \cdot \mathbf{B}^0 = (a_1 - \bar{a})(b_1 - \bar{b}) + (a_2 - \bar{a})(b_2 - \bar{b}) + (a_3 - \bar{a})(b_3 - \bar{b}) + (a_4 - \bar{a})(b_4 - \bar{b})$$

In general, in n-dimensional space

$$\mathbf{A}^0 \cdot \mathbf{B}^0 = \sum_{i=1}^{n} (a_i - \bar{a})(b_i - \bar{b})$$

If we divide this by $n - 1$, we obtain the well-known expression for *covariance*, which is used to establish the degree of association between two or more distributions. Covariance is calculated in the same way as *variance*, except that there are multiple distributions. The variance can be thought of as a measure of the

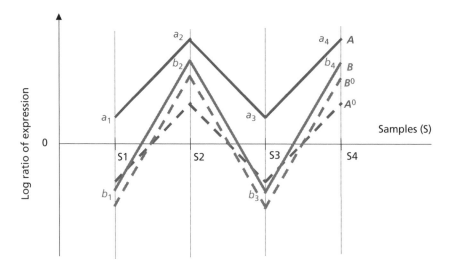

Figure 4.4 Mean centring (see text for explanation).

distance from the mean, or the 'spread' of the data. Covariance is the generalisation of variance for two distributions and can be expressed as

$$\text{Cov}(\mathbf{A}, \mathbf{B}) = \frac{\mathbf{A}^0 \cdot \mathbf{B}^0}{(n-1)}$$

The normalised covariance gives the expression for *linear correlation*, also known as the *Pearson correlation coefficient* (PCC):

$$\text{Cor}(\mathbf{A}, \mathbf{B}) = \frac{\mathbf{A}^0 \cdot \mathbf{B}^0}{|\mathbf{A}^0\|\mathbf{B}^0|}$$

In this way we see that the PCC between vectors **A** and **B** is the same as the angle distance between these vectors in normalised and mean centred space. For unrelated distributions the PCC is near 1 for a strong correlation and near zero for a weak correlation (Figure 4.5 and Plate 4.1, facing p. 88). We can compensate for different contributions from each column by dividing by the sum of the total expression in the column as we did for the chi-square distance. The angle may be larger than $\pi/2$ and the angle distance $\cos\alpha$ may vary in the range from -1 to 1, in which case we may have a negative correlation, sometimes also called *anticorrelation*. Note that for mean centred vectors the PCC equals the dot product.

Since the distance should be positive (according to the definition of distance), we have two possibilities for defining the distance: either $D_{\text{angle1}} = 1 - \cos\alpha$, or $D_{\text{angle2}} = 1 - |\cos\alpha|$. In the first case a perfect correlation will give a distance of zero, while anticorrelation will give the maximum distance of 2. In the second case correlation and anticorrelation are treated equally, and perfectly correlated or anticorrelated profiles will both have zero distance.

Since centring the vector space and normalising the vector length to 1 substantially simplifies the calculation of the correlation coefficient (for mean centred vectors it is expressed by the dot product), it is tempting to transform the

(a)   (b)   (c)

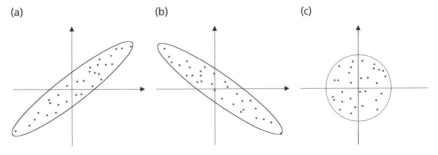

Figure 4.5 Pearson correlation. The linear correlations between two distributions, the values of which are represented by each axis, are positive in (a), negative in (b) and close to zero in (c).

original vector space (gene or condition) into the mean centred normalised space right at the beginning. In this space the chord distance becomes equal to the Euclidean distance and there is a simple relationship between the chord, the angle and the linear correlation distances (Figure 4.3).

It should be noted that transformation of the original vector space to such a 'normalised' space results in the loss of information – we can no longer tell how long the original vectors were, i.e. what the expression levels were. Although vector centring and normalisation preserves the relative 'shapes' of the expression profiles, the magnitudes are lost. If the expression levels were measured in absolute terms (e.g. if we are using the absolute fluorescence intensity values for analysis), such transformation can be justified only as a technical means to simplify calculation. On the other hand, if we use ratio or log ratio values, the absolute expression values may not be meaningful.

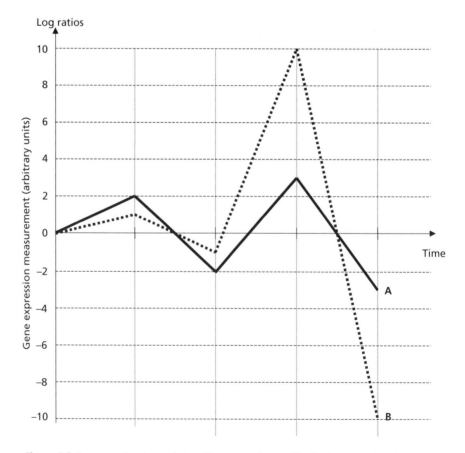

Figure 4.6 Pearson and rank correlation. Gene expression profiles for time series A and B.

4.2.3.3 SPEARMAN'S RANK CORRELATION

Consider two gene expression profiles for the time series $A = (0, 2, -2, 3, -3)$ and $B = (0, 1, -1, 10, -10)$, as shown in Figure 4.6. Note that the expression values are changing in a very similar way (always going up and down for the same time intervals). Both expression profiles are mean centred $(0 + 2 - 2 + 3 - 3 = 0$ and $0 + 1 - 1 + 10 - 10 = 0)$, therefore $A^0 = A$ and $B^0 = B$, and $A^0 \cdot B^0 = 0 \cdot 0 + 2 \cdot 1 + (-2) \cdot (-1) + 3 \cdot 10 + (-3) \cdot (-10) = 64$. Noting that $|A| = \sqrt{26}$ and $|B| = \sqrt{202}$ we obtain

$$\text{cor}(A, B) = \frac{64}{\sqrt{26 \cdot 202}} \approx 0.88$$

Although the correlation is strong, it is not perfect, which is intuitively understandable – the magnitude of the change in expression value is different between adjacent time points in series A and B.

Spearman's rank correlation (SRC) is a distance measure that is invariant to monotone changes, i.e. ignores the magnitude of the changes if their relative ranks within the series are preserved. The idea of the rank correlation is to transform the original values into ranks, and then to compute the linear correlation between the series of ranks.

More precisely, first we order the values in each of the series non-decreasingly and assign to each value a *rank* beginning with 1. For instance, the vector $A = (0, 2, -2, 3, -3)$ is transformed into rank vector $A_{\text{rank}} = (3, 4, 2, 5, 1)$ (the smallest value in A is -3, which gets rank 1, the second smallest is -2, which gets rank 2, . . . , the largest among the five elements of the vector has a value equal to 3, which gets rank 5). Similarly, $B = (0, 1, -1, 10, -10)$ is transformed to $B_{\text{rank}} = (3, 4, 2, 5, 1)$.

We define the rank correlation between A and B as the Pearson correlation between A_{rank} and B_{rank}. Note that $A_{\text{rank}} = B_{\text{rank}}$, therefore the correlation between them is perfect, i.e. equal to 1.

If two or more values in a series are repeated, the rank calculation is slightly more complicated – see, e.g. Francis (1988) – but for gene expression data this is not usually a problem.) A non-trivial derivation, which we do not present here, shows that the rank correlation coefficient can be expressed directly as

$$\text{Rank}(A, B) = 1 - \frac{6\left(\sum d_i^2\right)}{n(n^2 - 1)}$$

where d_i is the difference between the ranks at position $i = 1, \ldots, n$, and n is the number of dimensions, i.e. samples. An advantage of rank correlation is that it does not change by a large amount if there are outliers in the data. The disadvantage in using a rank correlation-based measure is that some information about gene expression levels may be lost in ranking the data.

4.2.3.4 DISTANCES IN DISCRETISED SPACE, AND MUTUAL INFORMATION

As already mentioned, sometimes it is advantageous to use a discretised expression matrix as the starting point, e.g. to assign values 0 (expression unchanged), 1 (increased) and −1 (decreased) as in Table 4.5. There are some distance or similarity measures that are defined only for discretised space. For instance, we can define the similarity between two discretised vectors as:

1 the number of positions that have the same value;

2 the number of positions that have the same non-zero value (i.e. we do not count zero for measuring the similarity);

3 the number of positions that have an equal non-zero value, divided by the total number or positions that have non-zero values in both vectors.

A useful similarity measure is based on the notion of *mutual information*. This can be defined for discretised as well as non-discretised matrices, but the definition is much simpler for the discretised matrix. The measure is based on the notion of *Shannon* or *information entropy*, and can be intuitively thought of as the amount of 'information' one can obtain from one expression profile, for predicting the behaviour of the other. This gives us a powerful similarity measure, which is more general than linear or rank correlation (although one should note that discretisation may result in a certain loss of information from the original profiles).

Shannon entropy characterises the level of uncertainty of a chosen position in the expression profile having a certain value. For instance, consider the discretised gene expression matrix in Table 4.7. If we consider the discretised values for gene A, they are all 'certain' to have value 1, while for gene C or D, the 'probability' of the value of a randomly chosen position being equal to −1, 0 or 1 is 1/3, therefore in a sense these profiles have the greatest possible 'uncertainty' (for the case of three possible values). Another way to see this is as an 'information gap' that needs to be filled in for knowing the (discretised) expression values of the gene. The values for gene B are more often equal to 1 than −1, and are never equal to 0, therefore they are not as 'uncertain' as the values for C and D.

Table 4.7 The expression levels for genes A, B, C and D have been discretised, to reflect the change in gene expression in relation to a reference sample.

A	1	1	1	1	1	1
B	1	1	1	−1	−1	1
C	1	0	−1	−1	1	0
D	0	1	−1	−1	0	1

The mutual information between two expression profiles is the decrease in 'uncertainty' (or 'information gap') of the values for one gene that can be obtained from knowing the values for the other gene. For instance, we can notice from the matrix that the expression values of genes C and D completely determine each other (if gene C has value 1, D has value 0, if C is 0, D is 1 and if C is −1, D is −1), thus from knowing one the uncertainty of the other is reduced to 0.

More precisely, let x be a variable that can assume discrete values $x_1, \ldots, x_k$ with relative frequencies $p_1, \ldots, p_k$. Assuming that $p_1 + \ldots + p_k = 1$ we can treat the frequencies as naïve probabilities. *Shannon entropy* $H(x)$ is defined as

$$H(x) = -\sum_{i=1}^{n} p_i \log p_i$$

In our case the gene expression values in the table are equal to −1, 0 or 1, with probabilities depending on the frequency of their occurrence for the particular gene. For instance, for gene A the probabilities of values −1 and 0 are equal to 0, while the probability of the value 1 is equal to 1. It is easy to calculate that $H(A)$ = 0. For genes C and D, $H(C) = H(D) = -3(1/3)\log(1/3) = \log 3$. For gene B the entropy equals $H(B) = -(2/3) \log(2/3) - (1/3) \log(1/3) = \log 3 - (2/3) \log 2$. We see that the entropy of C and D is maximal, the entropy of A is minimal, and the entropy of B is in between.

Having two variables x and y ranging over the set of discrete values (x_i, y_j) with frequencies p_{ij} we can define the joint entropy $H(x, y)$ as

$$H(x, y) = \sum_{i,j} p_{ij} \log p_{ij}$$

For instance, the joint entropy of C and D equals $H(C, D) = -3(1/3)\log(1/3) = \log 3$, since there are three possible values $(1, 0)$, $(0, 1)$ and $(-1, -1)$ (i.e. the 'probabilities' of all other pairs are 0), each with the same 'probability' $1/3$.

We define the *conditional entropy* $H(x|y)$ of x given y as

$$H(x|y) = H(x) - H(x, y)$$

The conditional entropy characterises the 'uncertainty' of random variable x, provided that we know the value of y. Note that for genes C and D, the conditional entropy $H(C|D) = \log 3 - \log 3 = 0$, which is intuitive, since knowing the discretised expression values of D, we can determine that of C.

The symmetric version of this is what is called *mutual information* and is defined as

$$M(x, y) = H(x) + H(y) - H(x, y) = H(x|y) + H(y)$$

and can be used as a similarity measure between x and y.

This ends our discussion of distance measures. No fewer than 26 different distance and similarity measures are listed in Legendre and Legendre (1998). It is not possible to provide a 'recipe' that describes which distance measure to use in each situation – one has to consider the nature of the data and the goals of the analysis, and then choose one or a few different measures. Variations of correlation coefficients and chi-square distances are probably the most commonly used ones. Although these measures are intuitive, they are not necesarily appropriate in all situations.

4.2.4 Principal component analysis, eigen-vectors and eigen-genes

Principal component analysis (PCA) is one of the most common methods used for gene expression data analysis, primarily to reduce the dimensionality of data and to find combinations of experiments or genes that jointly contribute most to variability in the data. Although the techniques used in PCA are not simple, the underlying idea is quite intuitive – it is based on finding the directions in multidimensional vector space that have the largest amplitude in the dispersion of data points. These directions then serve as new coordinate axes. It turns out that for many microarray datasets most of the variability can be accounted for by a small number of principal directions. By taking only two or three most important (most variable) directions one can visualise the data in two or three dimensions often without losing much information.

We start by introducing the most basic elements of linear algebra that are needed to understand PCA. Linear algebra is a rich and non-trivial subject; for a more comprehensive introduction see, for instance, Strang (1993). Full understanding of this subsection is not essential for being able to comprehend what follows in the text with the exception of a few, inessential, details, therefore the reader who has had enough of mathematics at this point may want to skip the rest of this subsection.

Given a vector $X = (x_1, \ldots, x_m)$, and a constant c, we define the *scalar multiplication* as a new vector $cX = (cx_1, \ldots, cx_n)$. A vector Y is said to be a *linear combination* of vectors $X_1, \ldots, X_n$ if it can be expressed as

$$Y = c_1 X_1 + \ldots c_n X_n$$

for some constants $c_1, \ldots, c_n$. If a row (or a column) in a matrix can be expressed as a linear combination of other rows (respectively, columns), then it is said to be linearly dependent on the other rows (respectively, columns). It is an important property of matrices, that the maximal number of linearly independent rows and columns is limited by the minimum of the number of rows or columns in the matrix. For gene expression matrices we often have thousands of genes (rows) but only tens of experimental conditions (columns), which effectively means that the behaviour of most genes can be explained via the small number of combinations of gene expression profiles.

We saw in the previous sections that vector space can be transformed into a new space by shifting the coordinate system to a new zero point. A more general transformation is a rotation of the coordinate system (e.g. Figure 4.7).

It can be shown that for coordinate system transformations that are based on a shift and rotation, the vector coordinates in the new (transformed) space can be expressed as linear combinations of the vectors in the original space (to be more precise, in the formal language of linear algebra the transformations involving shift are called *affine*, as they include adding a constant vector).

We can use linear combinations of vectors (rows or columns) to transform the original vector space into a new vector space so that the axes in the new co-ordinate system are oriented along the directions of greatest variability. This is the essence of principal component analysis: transformation of the original vector space to a new space by rotating the coordinate system so that the new axes are in the directions of greatest variability. An example is shown in Figure 4.7.

In general there may be as many principal components as there are linearly independent rows or columns in the matrix, although only a few 'largest' components have intuitive meaning. The technique used for finding the vector space where the axes are oriented in the direction of the principal components (we can call these *principal axes*) is called *singular value decomposition* (SVD). The

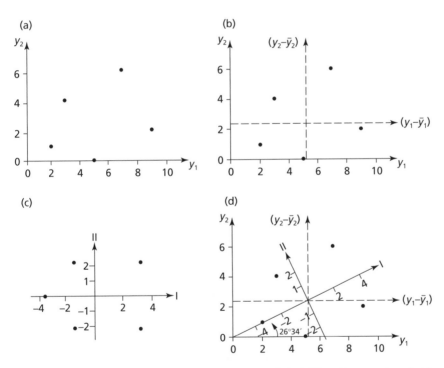

Figure 4.7 A numerical example of principal component analysis. (a) Five objects are plotted with respect to descriptors y_1 and y_2. (b) After centring of the data, the objects are now plotted with respect to $(y_1 - \bar{y}_1)$ and $(y_2 - \bar{y}_2)$, represented by dashed axes. (c) The objects are plotted with reference to principal axes I and II, which are centred with respect to the scatter of points. (d) The two systems of axes (b and c) can be superimposed after a rotation of 26°34′. (Reprinted from *Numerical Ecology*, 2nd edn, P. Legendre & L. Legendre, fig. 9.2, Copyright 1998, with permission from Elsevier Science.)

vectors representing the directions of the new axes are called *eigen-vectors*. For gene expression matrices, they are sometimes called *eigen-genes* or *eigen-conditions*. Note that the maximum number of eigen-genes in a matrix is no larger than the minimum number of experiments or genes, whichever is the smallest. Eigen-genes are mutually *orthogonal*, meaning that their dot products are equal to zero. This can be interpreted as zero correlation between the expression values on the principal axes.

Singular value decomposition can reveal different characteristics of the gene expression matrix. For instance, it may reveal which experiments (i.e. experimental conditions) provide the most significant contribution to the gene expression profiles, or which experiments are mutually correlated. It has been shown by Alter *et al.* (2000) that in many gene expression experiments only a few principal axes contribute to most of the variability in expression data.

One of the applications of singular value decomposition is visualisation of higher dimensional space in two or three dimensions. By finding the principal components we can project the *n*-dimensional vectors on the two (or three) dimensions that have the greatest variability. This is used, for instance, in correspondence analysis, discussed later. Where these two or three dimensions account for most of the variability in the data, the relative distances among points in the new space will reflect the distances in the original space. However, if there are more than two or three axes of high variability, the visualisation will distort the distances between data points in the original multidimensional data. The extent to which the distances in the original space are preserved in the reduced space can be used as an indicator of how appropriate the space reduction is. Space reduction in normalised, mean centred space (i.e. using a covariance matrix) is also known as *multidimensional scaling*.

4.2.5 Dealing with missing values

So far we have assumed that all the values in the gene expression matrix are defined, i.e. all expression levels are represented by real values. Unfortunately, in practice microarray data frequently contain missing values, i.e. some of the values in the matrix are undefined. Not all data analysis methods are directly defined for matrices with missing values. For instance, the distance measures described above are not directly applicable for vectors with missing values, although in many cases they can be generalised to include missing values in a straightforward manner (e.g. the dimensions containing missing values can simply be ignored). However, for some methods such a generalisation is more difficult and we either have to substitute the missing values by arbitrary values (e.g. zero) or try to estimate the missing values from the rest of the data.

Suppose the value x_{ij} for the gene i in the experiment j is missing. Three different methods that could be used for imputing this value from information in the existing data are listed below.

1 Row average method – simply uses the average value of the gene expression data from the gene of interest from other samples or other genes.

2 Weighted K-nearest neighbour method – finds K genes with the most similar expression profiles to gene i for which the values $x_{1j}, \ldots, x_{kj}$ are defined. The

method estimates the missing value x_{ij} from the values $x_{1j}, \ldots, x_{kj}$ using a weighted average, where the contribution of each gene is weighted based on how similar its expression profile is to that of gene i.

3 A method based on principal component analysis, more precisely: (i) finding the first k principal axes (as outlined in the previous section), i.e. the k most significant eigen-genes; (ii) projecting these eigen-genes onto the gene i and calculating the contribution of each eigen-gene; and (iii) reconstructing the missing value x_{ij} from the jth components of the eigen-genes, taking each proportional to the contribution of the respective eigen-gene to gene i.

These methods are described in detail in Troyanskaya *et al.* (2001). Evaluation comparing the three methods concluded that the K-nearest neighbour approach performed the best and that the optimum value of K ranged from 10 to 20. It should be noted that the weighted K-nearest neighbour approach has implications for downstream analysis, since gene expression data are frequently analysed with the objective of identifying genes with similar expression profiles.

4.2.6 Representation of gene expression data by graphs (networks)

Gene expression data may provide information about various relationships between genes, which often can be viewed as networks or graphs. We can define a *gene expression graph* or *network* as a graph where *nodes* represent genes and *edges* represent relationships between genes; an example is shown in Figure 4.8. For instance, we can draw an edge between genes A and B, if the correlation distance between the expression profiles of these genes is smaller than a predefined significance threshold. Note that such a graph is simply a representation of a discretised distance matrix, called an *adjacency matrix*, where 1 at the position (i, j) represents the edge between gene g_i and g_j. In fact any distance matrix can be transformed into a graph by applying a significance threshold, i.e. by discretising the data.

Labels can be attached to nodes or edges, in which case we talk about a *labelled graph*. For instance, genes can be labelled with their names or functional classes, and edges by the distances between the respective gene expression profiles. Edges can be oriented, representing a possible asymmetry in the relationship between the genes, in which case they are often called *arcs*. For instance, in the network represented in Figure 4.8, the grey arc from gene *SST2* to *FUS2* means that the expression of gene *FUS2* is greater in a strain in which *SST2* is mutated than in a non-mutant strain (Rung *et al.* 2002). On the other hand, the dashed arc from gene *FUS3* to *URA3* means that the expression of the gene *URA3* is greater in a non-mutant strain than in one where *FUS3* is mutated.

4.2.7 Gene expression matrix annotation

So far we have considered gene expression matrices separately from gene and sample annotation. The use of annotation has two applications. First, each gene or sample should be assigned a unique (standard) identifier, allowing re-

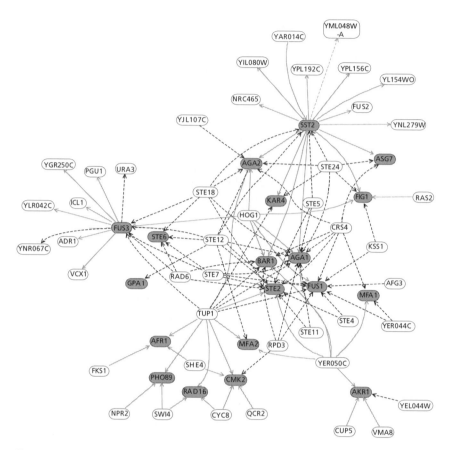

Figure 4.8 An example of a gene expression graph or network, where nodes (circles) represent genes and edges (arrows connecting the circles) represent relationships between genes.

searchers from different laboratories to communicate unambiguously (i.e. to make sure that indeed the same gene or experimental condition has been identified). Note that the lack of standards even for gene names is a serious problem. Second, the annotation can represent various types of information known about the particular gene (e.g. functional role) or experimental condition (e.g. particular disease state).

Annotation can either be in free text format or use controlled vocabularies (CV). If we use CVs, it is possible to incorporate the annotation in the mathematical analysis of the data by treating it as a categorical variable. Controlled vocabularies can either be *flat* (i.e. provided as a simple list of terms without structure) or organised. The simplest way of organising a vocabulary is hierarchical, i.e. in the form of a tree. An example is a species taxonomy tree. In a hierarchical CV every term can have only one parent term, but many descendent terms. Sometimes there is a need to make this structure more flexible and to allow the possibility of more than one parent. In this case we can organise the vocabulary in what is known as a *directed acyclic graph*, or DAG.

A system of controlled vocabularies for annotation of gene functions is being developed by the Gene Ontology consortium (GO) (see GO consortium, 2000 and http://www.geneontology.org/), and GO terms are becoming a more commonly accepted standard. The GO system works at three levels for each gene product: *molecular function*, *biological process* and *cellular component*. All three controlled vocabularies are independent and each is organised in a DAG (a hierarchical organisation is not sufficient; for instance, in biological process ontology, 'DNA ligation' is a subclass of 'DNA repair' and 'DNA recombination'). GO terms are facilitating the interpretation of the results of gene expression analysis, e.g. by permitting the display of gene functions in a concise way with their expression profiles (Plate 4.2, facing p. 88). It has been shown that in many cases biological processes defined by GO correlate with expression profiles.

Sample and experimental condition annotation is less developed. Only the organism classification has a well-adopted standard – the species taxonomy database maintained at the National Center for Biotechnology Information. Standards for cell and tissue types, organism parts, developmental stages and disease states are under development. For some model organisms, e.g. mouse, such standards are emerging through the efforts of community databases, such as the mouse database at the Jackson Laboratory. To standardise the annotation for treatments is an even more daunting task. The Ontology Working Group of the Microarray Gene Expression Data Society is currently developing standards for sample annotation (see http://www.mged.org/ontology/).

A simple way to annotate samples is possible when we can categorise them into two or more classes, e.g. representing diseased or normal tissues. In this case we can assign, for instance, 0 to the normal tissues and 1 to diseased ones, which effectively extends each column in the matrix by one row. Distance measures for vectors that include categorical values can be defined and used in the downstream analysis. Similarly, we may assign value 0 or 1 to genes, depending on whether the gene belongs to a particular GO category or not. Since each gene may be associated with many GO terms, it is not always trivial to decide which part of the vocabulary is most appropriate for each case; often more than one term for a gene can be used, and we may use more than one column for gene annotation.

In this way, adding annotation to the expression matrix may be considered an extension of the original matrix. However, the nature of the values in the extended columns is different from that of the expression values. Supervised analysis can be defined as a method of matrix analysis that uses these additional columns or rows in a particular way. In unsupervised analysis we use the annotation only to interpret the data after the analysis is completed.

4.3 Clustering

The goal of gene expression data clustering is to group together genes or samples that have similar expression profiles. Clustering is currently the most popular method of gene expression matrix analysis. It can be useful for discovering 'types' of behaviour, for reducing the dimensionality of the data (allowing tens of thousands of genes to be represented by a few groups each containing genes that behave similarly), as well as for the detection of outliers in the data.

Clustering is one of the unsupervised approaches to data analysis, which can be used in the absence of *a priori* information, or when annotations are not considered in the analysis. Clustering is a well-established field and various clustering algorithms have been invented, many of which can be considered as 'classic'. These include *hierarchical agglomerative clustering*, which is based on iteratively grouping together the objects that are most similar to each other, and *K-means clustering* (Hartigan, 1975), in which the number of clusters is defined *a priori*, and the clustering is iteratively improved by adjusting the cluster centres in Euclidean space. There are also newer clustering algorithms, such as Kohonen's self-organising maps (Kohonen, 1990), which is similar to K-means methods, graph theory-based algorithms (Sharan and Shamir, 2000), and methods that use PCA (Hastie *et al.*, 2000). Jain *et al.* (1999) give an overview of clustering algorithms.

Many of the well-known clustering algorithms have been used for clustering of expression profiles, including hierarchical clustering (Alon *et al.*, 1999; Eisen *et al.*, 1998), K-means clustering (Tavazoie *et al.*, 1999; Vilo *et al.*, 2000) and self-organising maps (Tamayo *et al.*, 1999; Törönen *et al.*, 1999). A simple clustering algorithm based on binning, i.e. discretising the expression profile space and clustering together the profiles that map to the same bin, has been shown to be useful for grouping data in situations where the number of experimental conditions is relatively small (Brazma *et al.*, 1998).

There is no compelling evidence that more sophisticated clustering algorithms perform better than the simplest ones with respect to the biological insights that have been obtained. At the same time this does not mean that all clustering algorithms are appropriate for all datasets. It is possible that in the future, when we have better understanding of the nature of gene expression data, more specialised algorithms that perform significantly better than others will be developed.

Before choosing a clustering algorithm, one has to think about which distance measure is likely to be the most appropriate for the particular dataset. We have to think about what type of clusters might be expected in the data. For example, consider Figure 4.10, where different clustering methods are likely to group these points in rather different ways.

Some clustering algorithms are applicable only for particular distance measures, e.g. K-means assumes that the distance measure has Euclidean properties. Another consideration is the performance of the various algorithms. When clustering thousands of genes for hundreds of experimental conditions, we are dealing with large data matrices, and some of the early implementations of clustering algorithms are not able to process these.

4.3.1 Types of clustering

Clustering can be either hierarchical or flat, as well as agglomerative or divisive. Agglomerative processes start out by considering each object as a separate cluster and proceed to group the most similar objects in an iterative fashion until all the data are included. Divisive methods start out with the complete set of data as one large group, or cluster, and proceed by partitioning the objects starting with those that are most dissimilar (Alon *et al.*, 1999). In addition to agglomerative or divisive methods, the clustering algorithm may start with the partitioning of the data into the predefined number of clusters, and then refine the assignment of the objects to clusters by changing the cluster boundaries. *K*-means is an example of such an algorithm.

Hierarchical clustering algorithms typically build a tree (also known as a dendrogram; see below) that represents a hierarchical structure in the data (e.g. see Figure 4.11 and Plate 4.3, facing p. 88). Flat clustering gives either a partitioning of the object space into a number of subsets (Figure 4.9a) or a system of overlapping clusters, which are not organised in any particular hierarchy (e.g. Venn diagrams – Figure 4.9b). A typical example of a flat partitioning is *K*-means (see Section 4.3.4). The number of clusters in a flat clustering can either be provided by the user or determined based on various theoretical or heuristic principles.

A different way to categorise clustering algorithms is based on whether the algorithm optimises a defined scoring function rating the 'goodness' of the

(a)   (b)

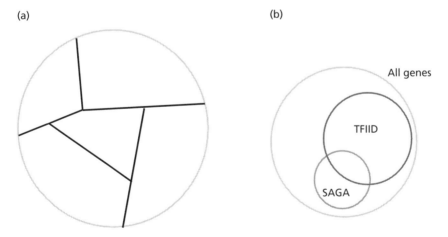

Figure 4.9 (a) A polygonal diagram and (b) a Venn diagram. (TFIID and SAGA are multi-subunit complexes that are involved in the transcription of protein-encoding genes. The two smaller circles represent the genes affected by mutations in individual subunits of TFIID and SAGA respectively.) ((b) reprinted with permission from *Nature* (Lee *et al.*), Copyright 2000 Macmillan Magazines Ltd.)

Figure 4.10 Which are the clusters?

clustering, or is based on some heuristics such as iterative joining of the most similar objects. Most of the commonly used clustering algorithms are heuristic. It is not easy to define a good optimisation function rating the 'goodness' of clustering. For instance, consider Figure 4.10: Are there any clusters in these data? How one would describe them? Even answering the question, 'How do we find the best balance between the cluster size and the compactness?' is not a trivial problem. Some possible cluster optimisation functions are described later in this chapter.

4.3.2 Hierarchical agglomerative clustering

Hierarchical agglomerative clustering is a process in which the data are successively fused, typically until all the data points are included. For hierarchical agglomerative clustering usually all the pair-wise distances between objects need to be defined. An agglomerative process typically starts by considering each object/data point as a separate, or singleton, cluster. Starting with n objects, the result of the first iteration of clustering is that the two objects that are most similar are grouped together to form a single cluster, leaving $(n-1)$ clusters. The distance between the objects and the newly formed cluster containing two objects is then updated and the next most similar objects and clusters are grouped together as a single cluster (Eisen *et al.*, 1998). This process is carried out iteratively until there is a single large cluster, as shown in Figure 4.11 for five data points.

The results of hierarchical clustering are frequently represented in a hierarchical tree, also known as a *dendrogram* (Figure 4.11 and Plate 4.3, facing p. 88). The branch lengths of the tree may represent the degree of similarity between the data. Note that the threshold values, i.e. where to split the tree to obtain distinct clusters, have to be established independently. Usually this is done quite arbitrarily by picking the clusters either that seem to be the tightest, or that include genes with similar functional annotation.

The process of hierarchical clustering illustrated in Figure 4.11 uses the minimum distance for determining which of the objects and clusters are most similar. There are, however, several methods by which the distance between the

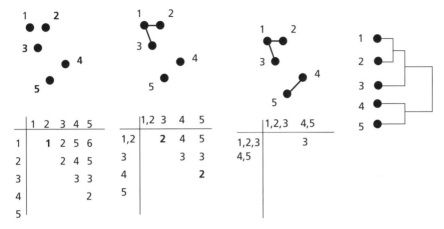

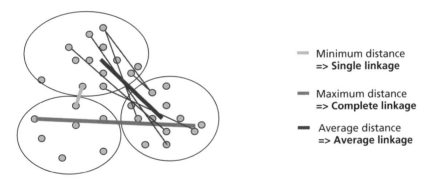

Figure 4.11 Steps in hierarchical clustering. Objects with the minimum distance between them are grouped together to form a new cluster, or 'single' object. The distance between all the clusters, or objects, is recalculated and the cycle is repeated iteratively.

Figure 4.12 Single, complete and average linkage.

clusters – or between clusters and objects – can be measured (summarised in Figure 4.12):

1 Single linkage (also known as nearest neighbour linkage, not to be confused with the supervised approach known as nearest neighbour analysis) uses the minimum distance between objects in the two clusters as the measurement of the distance between the clusters. This results in clusters that are spread out, as the grouping is based on minimum dissimilarity between members of each group. The method is not sensitive to outliers.

2 Complete linkage clustering groups objects according to the greatest distance between the objects in the clusters. Complete linkage is also known as the maximum or furthest neighbourhood method. Complete linkage tends to form tight clusters of similar objects as the distance between objects is based on

the greatest dissimilarity between them. A disadvantage of this method is that it is sensitive to outliers.

3 Average linkage clustering measures the distance as the average distance between every point in a cluster and every point in the other cluster. The average can be calculated in a number of ways. Average linkage methods can be either weighted or unweighted, depending on whether we compensate for the size of the cluster, or treat clusters of different sizes equally. When using a clustering algorithm based on average linkage it is useful to know how the average is calculated.

4 The centroid method is similar to the average linkage method, the difference being that it uses the 'mean centroid' of each cluster to represent the cluster as a whole. The 'centroid' is calculated by taking the mean for each dimension separately. (In Euclidean space the centroid can be visualised as the intuitive centre of gravity of the cluster.) The advantage of the centroid method is that the properties of each cluster are represented by one object – a centroid. The disadvantage is that the centroid position and the distances from all centroids to the new one have to be recalculated with every merging of the clusters, which affects the performance of the algorithm (for non-centroid methods the distances can be updated more efficiently). Centroid clustering may cause 'reversals' in the dendrogram (Figure 4.13), since the distance to the centroid of a new cluster C resulting from merging clusters A and B may be smaller than that between A and B. In practice this does not cause problems, as the reversals are typically rather small. Like average linkage clustering, centroid clustering can be weighted.

5 In Ward's method, also known as minimum variance clustering, the distance between objects and clusters is obtained by calculating the sum of the squared distances from the mean centroid of each cluster. This creates a measure of dispersion or of heterogeneity between objects in the group. Objects A and B are joined only if the increase in the sum of the squared distances between A and B

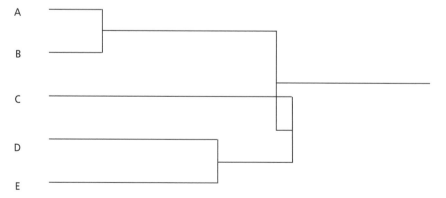

Figure 4.13 Centroid clustering may cause 'reversals' in the dendrogram, e.g. if the distance to the centroid after merging A and B is smaller than the distance between A and B.

is less than the increase in the sum of the squared distances of joining other objects. Ward's method favours small, equally sized clusters, with minimal within-group dispersion; however, it is computationally expensive.

While intuitively appealing as a method, hierarchical clustering has some limitations. The process of grouping continues until all the clusters are joined, and therefore in the end objects that have no similarity to each other are grouped together. This means that, in practice, the most relevant groupings are those that relate small numbers of genes. Each iteration of the cycle produces a fixed classification that is built upon in successive iterations of the algorithm; there is no opportunity to re-evaluate the groupings that were assigned early. This makes hierarchical methods less robust, i.e. small changes in the data can produce a completely different clustering, and as a result hierarchical clustering is less suitable for noisy data.

The disadvantage of a 'classical' hierarchical clustering is also that the full distance matrix of all pair-wise distances has to be calculated in advance, which for n objects takes on the order of n^2 steps. For large gene expression data matrices consisting of tens of thousands of genes, the performance of the algorithm may prove a limitation. An alternative approach to hierarchical clustering that does not require the calculation of all pair-wise distances is self-organising trees (see Section 4.3.5).

4.3.3 Hierarchical divisive clustering

Hierarchical divisive clustering starts with the whole set of objects and divides them into two or more subgroups. After this each subgroup is considered separately and the division is repeated iteratively (Lance and Williams, 1967).

Partitioning is usually done along one of the axes of the n-dimensional space at a time. Principal components analysis can be used to find the principal axes at each step. The division can be done either along the most significant original axis, or along the new axes in the transformed space. Various hierarchical divisive methods are described in Legendre and Legendre (1998). The method has been applied to gene expression analysis (Alon *et al.*, 1999), though overall it is less popular than hierarchical agglomerative methods.

Hierarchical clustering organises the data in a tree and leaves it to the user to define similarity thresholds that are used to determine which objects should be considered as a cluster. An advantage of this approach is its flexibility, which allows the user to choose the particular clusters from the dendrogram. This is also a disadvantage, however, as it makes fully automated data analysis more difficult. Hierarchical clustering has been criticised by those who argue that there are no compelling reasons why there should be any 'real' hierarchical structure in gene expression data. The popularity of hierarchical clustering in biology stems partly from the successful application of hierarchical taxonomic principles in many different contexts. Such relationships may not exist in gene expression data; nevertheless, hierarchical clustering provides a powerful tool for organising thousands of expression profiles in a way that has been proved to be useful.

4.3.4 Non-hierarchical clustering – *K*-means

Non-hierarchical, or flat, clustering groups the data into non-overlapping, or overlapping, clusters. The number of clusters can either be given in advance or estimated from the data by applying various criteria. The most common non-hierarchical methods are *K*-means clustering, self-organising maps and various so-called Bayesian approaches. Self-organising maps impose a partial structure on the data, such that adjacent clusters are related. Bayesian clustering permits the incorporation of priors, i.e. additional information about our knowledge of the data.

K-means is the most common method of partition-based clustering. It starts with the given number of cluster centres, chosen either randomly or by applying some heuristics, some of which are described below. Next the distance from the centroids to every object is calculated, and each object is assigned to the cluster defined by the closest centroid; then, for each cluster the new centroid is found. The distance from each object to each of the new centroids is calculated and in this way the boundaries of the partitioning are revised. This is repeated either until the centroids stabilise (which is not guaranteed) or until an *a priori* defined maximum number of iterations has been reached. It is common that algorithms proceed through 20,000 to 100,000 cycles before the position of the nodes stabilises. This process is illustrated in Figure 4.14.

The initial position of the centroid nodes, called *seeds*, can be determined by one of several methods. A random selection of objects can be used to represent

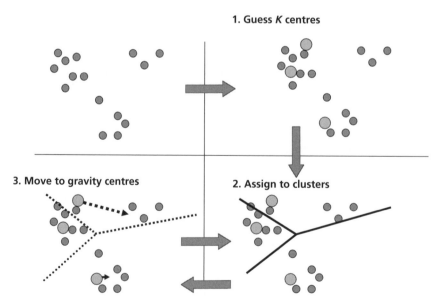

Figure 4.14 *K*-means clustering (see text for explanation). The large circles represent the centroids and the small circles represent the objects.

each of the nodes (if this method is used, it is usually carried out a number of times, i.e. multiple seeds are used). The nodes can be chosen deterministically by selecting the objects most distant from the gravity center of the whole data as well as from each other. Another possibility is to choose the centres of K nodes after hierarchical clustering of a small subset of objects (Li and Vitanyi, 1993).

As the initial position of the nodes may influence the final partitioning, some implementations permit multiple 'rounds of clustering'. For instance, clustering is carried out three times starting with different seeds each time and the most stable clusters, or those that best meet a criterion, such as minimisation of within-cluster variance, are used as the final result. The optimum number of seeds is such that the assignment of genes to clusters and the expression profile of each cluster is stable.

Finally, it should be noted that K-means is one of the most scalable algorithms for large datasets.

4.3.5 Self-organising maps and trees

Self-organising maps were first introduced by Kohonen in 1990 (Kohonen, 1990) and were first used to analyse gene expression data by Tamayo *et al.* (1999). The method has since been implemented in a number of gene expression analysis programs (Tamayo *et al.*, 1999; Törönen *et al.*, 1999). Self-organising maps (SOM) work well with noisy data (Mangiameli *et al.*, 1996). Self-organising map-based algorithms are a divisive clustering approach. Prior to initiating the analysis, the user defines a geometric configuration for the partitions, typically a two-dimensional rectangular or hexagonal grid, and the number of clusters. Each cluster is represented by a node called a 'reference vector' and the reference vectors are placed on the chosen grid.

Let us consider the situation shown in Figure 4.15, where we have 19 objects in expression space. Suppose we have decided on the rectangular geometry and the number of clusters as equal to six. The nodes are projected onto the gene expression space and each of the data points is assigned to the nearest node, in a process known as initialisation. After the initialisation the following two steps are iterated:

1 A gene is picked at random.
2 The reference vector that is closest to the selected gene is moved closer to the randomly picked gene.
The reference vectors that are nearby on the two-dimensional grid are also adjusted by small amounts so that they too are more similar to the randomly selected gene. Increasing the stringency is used to define closeness in each step.

Finally, the genes are mapped to the relevant partitions depending on the reference vector to which they are most similar.

In choosing the geometric configuration for the clusters, the user is, effectively, specifying the interrelationships between clusters. As with K-means clustering, the user has to rely on some other source of information (possibly empirical) to determine the number of clusters that best represent the available data.

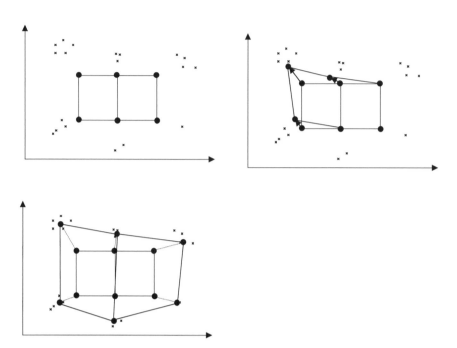

Figure 4.15 Self-organising maps (see text for explanation). The initial positions of the reference vectors are depicted by the shaded circles.

The self-organising tree algorithm (SOTA) is one of the newer methods and was developed by Dopazo *et al.* (Dopazo and Carazo, 1997; Herrero *et al.*, 2001). The algorithm combines ideas taken from hierarchical clustering and self-organising maps to produce clustering that is claimed to be more robust than classical hierarchical clustering, as well as more efficient. The SOTA produces a hierarchical, binary organisation of the data and proceeds divisively. A criterion, based on comparison of the data with a randomised sample of the same data, is provided as statistical support for the clustering obtained, and used to assist in calculating the point at which the data should no longer be sub-divided into individual clusters. The 'growing' process may be terminated when the heterogeneity of the gene expression profiles in a cluster falls below a predefined threshold, giving rise to asymmetric growth, or may proceed until all the data are incorporated into the hierarchical structure, producing a result similar to that obtained by classical hierarchical clustering.

4.3.6 Relationship between clustering and PCA

Both clustering and principal component analysis can be considered a means of reducing dimensionality in the data, though in rather different ways. Clustering groups expression profiles by similarity and each group can then be considered as a separate object, e.g. an object represented by a centroid. In this way, the

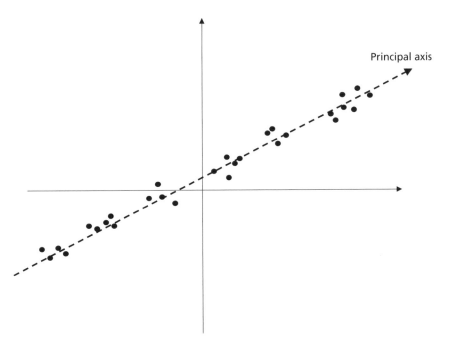

Figure 4.16 There is one principal axis describing most of the variability in the data, but no fewer than six distinct clusters.

original gene space of possibly thousands of genes can be reduced to a few clusters.

Principal component analysis, on the other hand, can be used to find the directions in the data space which account for the greatest variability in the data, orient the coordinate axes in these directions, and recalculate the expression profiles in the new transformed space. If only a few directions account for most of the variability (which is often a property of gene expression datasets), then the other directions are less important, and in this way the dimensionality is reduced to the few important axes. These axes can sometimes be interpreted as a combination of experimental conditions.

Note that although both methods reduce the dimensionality, they are entirely different, and PCA should not be confused with clustering. For instance, the data may have only one important axis accounting for, say, 90% of the variability, while at the same time there may be many distinct clusters in the data arranged along this axis (e.g. as in Figure 4.16).

4.3.7 'Gene shaving'

Gene shaving is a clustering method, introduced by Tibshirani *et al.* (1999), that exploits PCA. This clustering method relies on an iterative heuristic algorithm that identifies subsets of genes with coherent expression patterns and large variation across conditions. The algorithm works iteratively by alternating

PCA analysis and picking the 'best' clusters consisting of genes contributing most to the variability in the data.

The algorithm starts by finding the first principal component in gene space. Then starting from genes that are most similar to the first principal component, the algorithm builds a mutually inclusive system of clusters of various sizes including these genes. For each of the inclusive clusters it estimates the 'quality' of the cluster using a measure called a 'gap statistic' (a measure introduced by the authors – for details see the original paper) and chooses the cluster with the highest score. Next, the algorithm transforms the gene space by removing the component from each gene that is parallel to the first principal axis (this is known as *orthogonalisation* of the data with respect to the first principal component). The algorithm iterates the process.

Gene shaving is different from the methods discussed above in that it does not produce a partitioning – a gene may belong to several clusters.

4.3.8 Clustering in discretised space

Discretisation of the gene expression data matrix allows one to use several clustering methods that are difficult or impossible to define for a continuous space. Remember that discretisation allowed us to define some similarity measures that are not defined for continuous space. These can also be used in clustering.

Discretisation divides the space into a finite number of bins, therefore discretisation itself can be used as a way of clustering – each expression profile from the original space maps into one of the bins, and each bin that is not empty (i.e. contains at least one expression profile) can be considered a cluster. For instance, if we have two experimental conditions, and the gene expression levels are discretised into three groups -1, 0 and $+1$, then each expression profile can fall into one of the nine bins: $(-1, -1), (-1, 0), \ldots, (+1, +1)$, where the first component is the discretised expression level of a gene for the first experimental condition, and the second component is the second experimental condition. The use of this simple clustering method has been described in Brazma *et al.* (1998). Essentially the same method has recently been applied for clustering gene expression time course data (Filkov *et al.*, 2001).

One problem with binning methods is that the number of bins grows exponentially with the number of experimental conditions, and if a considerable part of them are populated, there may be too many clusters for the clustering to be useful. A way round this is to perform divisive hierarchical clustering component-by-component until the desired number of clusters has been reached.

4.3.9 Graph-based clustering

Formally, graph theory defines a *graph* as a set of *nodes* (sometimes also called *vertices*) and a set of *edges*, i.e. $G = (N, E)$, where N is the set of nodes and E the set of edges (see, e.g., Harary, 1969). A *subgraph* is a graph that consists of a subset of the nodes of the original graph and all the edges in the original graph that are adjacent to the nodes in the subgraph. A subgraph is called a *clique* if it contains all the edges between any pair of the nodes of the subgraph. We can also talk about 'approximate cliques' i.e. subgraphs that can be transformed into cliques by adding a small number of edges. A graph is *connected* if for every pair of the nodes in the graph there is a path from one node to the other following the edges. If a graph is used to describe relationships between genes, the connectivity in the graph can be used to find modules in the underlying network (approximate cliques can be thought of as clusters). A novel graph theory-based clustering algorithm CLICK, based on a representation of the gene expression matrix as a graph (see Section 4.2.6) and designed specifically for gene expression profile clustering, has been proposed (Sharan and Shamir, 2000). This algorithm does not require any prior assumptions about the structure or the number of clusters.

The initial graph is defined by edges connecting two genes in the graph if the distance between these genes (e.g. correlation distance) is below a particular threshold. Each edge is weighted according to the similarity between the genes. The underlying assumption is that the 'real' clusters correspond to approximate cliques in the graph. The CLICK algorithm looks for the approximate cliques. In each step the algorithm considers a particular connected component of a subgraph consisting of yet-unclustered objects. If this component satisfies a particular 'tightness' criterion, it is declared a *kernel*, otherwise it is split into two, according to a so-called *minimum weight cut* (minimum weight cut is defined in graph theory, and intuitively means the set of edges with the minimum total width, the removal of which means that the graph would split into two). Once the graph is split into a number of kernels, each kernel is expanded into a cluster by adding the closest singleton objects (i.e. nodes that were not originally connected to any of the kernels). For full details of this algorithm see the original paper (Sharan and Shamir, 2000).

4.3.10 Bayesian or model-based clustering and fuzzy clustering

Most of the clustering methods described above are heuristic in the sense that they do not try to optimise any scoring function describing the overall quality of the clustering. Model-based clustering assumes that the data have been generated by some, typically probabilistic (Bayesian), model, and tries to find the clustering corresponding to the most probable model. The methods may still be heuristic in that they may not guarantee identification of the most probable clustering.

Various model-based approaches have recently been introduced. Although model-based clustering has the potential to incorporate *a priori* knowledge about the domain in the analysis, it is not easy to apply it in a way that produces more meaningful biological results than purely heuristic methods.

The clustering methods discussed so far are all deterministic, by which we mean that given a cluster and an object, the object either belongs or does not belong to the given cluster. Fuzzy clustering assigns to each object the likelihood (or probability) of belonging to the particular cluster. Bayesian methods are often used for fuzzy clustering.

4.3.11 Clustering genes and samples – applications of clustering

Clustering can be used either for genes or for samples, or for both simultaneously. When applied to genes, clustering helps to identify genes that are co-regulated or that participate in similar biological processes. This can be used, for instance, for promoter prediction (Brazma *et al.*, 1998; DeRisi *et al.*, 1997; Spellman *et al.*, 1998; Vilo *et al.*, 2000) as well as for prediction of gene function.

Alizadeh *et al.* (2000) applied hierarchical clustering to samples, for clustering tumour samples based on their expression patterns, and used this to find new potential tumour subclasses. Diffuse large B-cell lymphoma (DLBCL) was studied using 96 samples of normal and malignant lymphocytes. Applying hierarchical clustering to these samples, Alizadeh *et al.* showed that there is diversity in gene expression among the tumours of DLBCL patients. The authors identified two molecularly distinct forms of DLBCL that have gene expression patterns indicative of different stages of B-cell differentiation. These two groups correlated well with patient survival rates, thus confirming that the clusters are meaningful and showing that gene expression data have prognostic value.

The sample clustering approach has been combined with gene clustering to identify which groups of genes are the most important for the sample clustering (Alizadeh *et al.*, 2000; Alon *et al.*, 1999). Alon *et al.* applied a partitioning-based clustering algorithm to study a gene expression matrix comprising 6500 genes in 40 tumour and 22 normal colon tissues. Clustering by both genes and samples, a method known as two-way clustering, permitted identification of groups of genes characteristic of each of the sample types.

Although general principles of gene and sample clustering are the same, the same distance measures and algorithms are not necessarily optimal for clustering of each. Note that in most cases gene expression matrices contain many more genes than samples. Therefore, when clustering samples, there are many more attributes (genes) for each object (sample) than the number of samples. The expression levels of most genes may be unimportant for clustering, but as a set they can influence the outcome of the clustering substantially. The clusters will be more robust if genes not important for the analysis are filtered out before clustering is performed. Usually the genes whose expression does not change substantially in more than a certain percentage of samples are excluded from cluster analysis.

Note also that while gene expression profiles can be compared and clustered even if the reference samples and experimental and normalisation protocols for different columns in the matrix are different, one should attempt to cluster samples only if the data for each sample have been generated consistently. To demonstrate this, let us consider the example given in Table 4.8. The right part

Table 4.8 The left part of the table shows the hypothetical absolute gene expression levels under conditions A (labelled red), A' (labelled green), B (labelled red) and B' (labelled green), and the mixture of extracts A' and B (labelled green). A and A' are replicates of each other, as are B and B'. The right part of the table shows log ratios.

	A	A'	B	B'	(A' + B)/2	log A/B	log B'/A'	log 2A/(A' + B)	log 2B'/(A' + B)
G1	5	6	1	1	3.5	2.32	−2.58	0.51	−1.81
G2	6	5	2	1	3.5	1.58	−2.32	0.78	−1.81
G3	11	12	3	2	7.4	1.87	−2.58	0.55	−1.91
G4	2	2	10	11	6.0	−2.32	2.46	−1.58	0.87
G5	1	2	11	10	6.5	−3.46	2.32	−2.70	0.62
G6	7	7	8	9	7.5	−0.19	0.36	−0.10	0.26
G7	4	5	4	4	4.5	0	−0.32	−0.17	−0.17

of the table represents a log ratio matrix obtained in four two-channel hybridisation experiments, for two experimental conditions A and B and seven genes G1, . . . , G7. The left part represents the 'real' gene expression levels (in abstract absolute units) under each of the experimental conditions. In reality we do not know these values – all we use is the ratios.

Assume that in the first hybridisation the condition A (e.g. labelled red) is compared directly with condition B (e.g. labelled green), and the result is represented as the red/green ratio A/B. In the second hybridisation a replicate of the condition A, denoted by A', is compared with a replicate of condition B, denoted by B'; the labelling is done reciprocally, therefore the result is represented by the red/green ratio B'/A'. In the third and fourth hybridisations, the samples A and B' are compared with an artificial reference sample made by mixing A' and B taken in equal amounts. The ratio is normalised (i.e. half of the total amount of A' + B is taken), and therefore the red/green ratio values are equal to 2A/(A' + B) and 2B'/(A' + B).

The absolute values of gene expression are given in the left part of the table, the ratios in the right part. If we consider the absolute levels of gene expression, the gene expression profiles roughly fall into three clusters: cluster 1: {G1, G2, G3} – high expression in condition A and low in B; cluster 2: {G4, G5} – low expression in condition A and high in B; and cluster 3: {G6, G7} – expression not changed substantially. Cluster 1 can be split into two subclusters depending on the absolute level of gene expression: {G1, G2} – lower absolute expression level; and {G3} – higher expression value.

Now let us consider the respective relative expression values given by log ratios in the right part of the table. Note that a gene clustering on this ratio table, i.e. clustering the rows, would reproduce the same essential clusters 1, 2 and

3 as the absolute expression values. It would not find the two subclusters of cluster 1; nevertheless, we can say that gene comparison is meaningful, even if the reference samples are different for each column.

The comparison between the sample expression profiles (i.e. columns in the matrix) is meaningful only if the reference sample is the same and the protocols standardised for all experimental conditions that we want to compare. For instance, clustering of the ratio columns in the table would be meaningless, while for the absolute measurements the replicates would cluster together as we expect.

4.3.12 Cluster scoring and validation

There is no simple answer to the question of how to find if there are any clusters in data, how many clusters there are, or how good a particular clustering is. It is often easier to devise a heuristic clustering method than to define a good scoring function characterising the overall 'goodness' or quality of the clustering. If the data contain well-defined clusters, a function that scores these clusters highly can usually be found. Unfortunately, in gene expression data there are often no strong clusters, either because of the presence of noise, or because of the nature of the particular underlying biological process. In these cases finding good optimisation functions is particularly difficult. However, if such a function can be defined, it can be used either for guiding the clustering or for assessing the 'goodness' of particular clusters or the clustering system after the clustering is complete.

To assess the cut-off thresholds and thus the number of clusters in hierarchical clustering, we can look for any 'jumps' in the distribution of distances used for merging the clusters. If a 'jump' is present in this distribution, this might be a natural cut-off point for defining separate clusters. Some clustering software packages (e.g. Expression Profiler) provide visualisation of this distance distribution, facilitating decision-making for the user.

Several measures and techniques to assess clustering quality have been proposed, but none of these is universally accepted. The most popular method for assessing the significance of the clustering and possible cut-off thresholds is based on randomisation of the data, e.g. by shuffling, which removes the correlations between individual data points, followed by comparison of the clustering with the shuffled data (Efron and Tibshirani, 1991; Tusher *et al.*, 2001). It is important to preserve the characteristics of the original dataset, including the range of values, frequency, and number of points, thus avoiding making assumptions about the data or the way in which the data are distributed. This approach can be used, for instance, with the self-organising tree algorithm to assign a threshold at which the data are considered sufficiently similar to stop subdividing them into separate clusters (Herrero *et al.*, 2001).

The 'goodness' of a cluster depends on how close its objects are to each other, and how far they are from the next closest cluster. A clustering significance measure based on this observation has been proposed by Rousseeuw (1987) and is called a *silhouette plot*. For each object in a cluster we define the average distance to the other objects of that particular cluster, as well as the average distance to the objects of the next closest cluster. In 'good' clusters the average distance within the cluster should be smaller than the distance to the next closest cluster for most objects.

Finally, let us note that the ultimate proof of the significance of clustering is whether it produces biologically meaningful results.

4.4 Classification algorithms and class prediction

Clustering algorithms try to find structure in the data without using any external information. Classification algorithms, by contrast, use external information, such as annotation, right from the beginning and try to find properties in the data that support this information. For instance, given a gene expression data matrix with samples annotated as 'diseased' or 'normal', the classification algorithm would look for combinations of genes that are expressed either only in diseased samples or only in normal ones. If such genes are found, the knowledge can later be used to assign 'diseased' or 'normal' states based on gene expression data only, i.e. for diagnostics.

Classification is a typical example of a supervised data analysis approach. The first well-known application of supervised learning to gene expression was classification of 72 leukaemia samples, from which 25 were acute myeloid leukaemias (AML) and 47 were acute lymphoblastic leukaemias (ALL), by Golub *et al.* (1999). Expression data from 6817 genes were used and expression profiles characteristic of each of these diseased states were found.

Classification algorithms can be used either to classify samples, e.g. to find specific expression patterns distinguishing various cell types, or to classify genes, e.g. to find specific expression patterns for genes in a particular GO category. There are examples of both applications in the literature, although in practice sample classification is more popular. There are at least two reasons for this. First, reliable classification of various diseased samples has a diagnostic value. Second, while there are many meaningful sample classifications that are entirely based on morphology, it is not easy to group genes into meaningful functional classes without using gene expression data in the first place.

There are many different classification algorithms, which have been developed mostly by the machine learning and statistical learning communities. The simplest of these are linear regression and K-nearest neighbour methods. One of the classification methods that is gaining popularity in gene expression analysis is support vector machines (see below). Typically, these classifiers are 'trained' on a subset of data with an *a priori*, given classification (the training set) and tested for accuracy on another subset with known classification (the test set). After assessment of the quality of the prediction, the classifier can be applied to new data. For example, in the above mentioned paper by Golub *et al.* the classifier was initially built from 38 leukaemia samples (11 AML and 27 ALL), and later tested on 34 samples, of which 29 were predicted accurately.

We will begin by formulating the classification problem more precisely.

4.4.1 Definition of the problem

Consider the $m \times n$ gene expression matrix $\mathbf{X}$ given in Section 4.2.2. Let us denote the columns of $\mathbf{X}$ by $A_1, \ldots, A_m$ (i.e. $A_1 = (x_{11}, \ldots, x_{n1}), \ldots, A_m = (x_{1m}, \ldots, x_{nm})$). Suppose each column A_i has a label $L_i \in \{1, -1\}$ (i.e. assuming values 1 or -1) attached to it, i.e. we have matrix

$$
\begin{pmatrix}
L_1 & L_2 & \cdots & L_m \\
x_{11} & x_{12} & \cdots & x_{1m} \\
x_{21} & x_{22} & \cdots & x_{2m} \\
\cdots & \cdots & \cdots & \cdots \\
x_{n1} & x_{n2} & \cdots & x_{nm}
\end{pmatrix}
$$

Vector $(L_1, \ldots, L_m)$ is called the *class vector* and labels L_i are called *class variables*. The domain of labels can be generalised to any controlled vocabulary. The set $(A_1, L_1), \ldots, (A_m, L_m)$ is called the set of *labelled profiles*.

A *classifier* is an algorithm which, given an expression profile A_i, outputs (predicts) the label L_i'. If for the given A_i the label is predicted correctly, i.e. if $L_i' = L_i$, we say that the profile is *classified correctly*. If on the other hand $L_i' \neq L_i$, we say that the profile has been *misclassified*. Given a classifier and labelled set $(A_1, L_1), \ldots, (A_m, L_m)$, we can count the number of labels that are incorrectly assigned and use this as the classification *performance estimator*. In fact this simple count is not a good performance estimator and in practice more sophisticated estimators (e.g. estimators that measure specificity, sensitivity or various combinations of both) are used, which we do not discuss here.

After we have chosen the type of classifier we want to use, we have to train it on the given labelled dataset with the goal of achieving the best prediction performance. We can use a subset of the data to *train* the classifier, and another subset to test the performance. In practice, when dealing with sample classification often we have datasets that are too small to leave out more than a few samples from the training set. Therefore we can use an approach that systematically leaves out one (or a given number of) profile(s), builds the classifier from the rest of the data, measures the performance of the unused data and repeats the process. The average misclassification count for all iterations can be used to rate the performance.

Different types of classifiers use different *training algorithms*. In fact, developing an efficient training algorithm is the most non-trivial part of building a classifier. Often, when talking about classifiers, in fact we mean the chosen type of a classifier together with the chosen training algorithm. The assumption underlying this approach is that if the dataset is representative and the classifier performs well, it will correctly classify a new expression profile A_{m+1} correctly. Machine learning theory studies the conditions under which this assumption is likely to be correct for different types of classifiers. The Occam's razor principle captures an empirical observation – that among the classifiers that correctly classify the given data, the simplest one has the best chance of performing correctly on new data.

One of the ways the Occam's razor principle is applied in practice is as the basis of the so-called *minimum description length* (MDL) principle (Li and Vitanyi, 1993). This principle notes that there may be a trade-off between the simplicity of a classifier and its accuracy. It assigns relative weights to the complexity of the classifier and the amount of misclassified data, and tries to minimise the sum.

When dealing with gene expression data, and particularly when trying to build sample classifiers, one problem is that there are usually many more attributes (genes) than objects (samples) to be classified: typically there are around ten thousand genes and fewer than a hundred samples. Therefore it is usually possible to find a few genes correctly classifying the samples, which at the same time may not give robust classification when the data are extended to new samples. For instance, for the above mentioned leukaemia example it turns out that using just three or even two first principal components in gene space (i.e. two or three eigen-genes built from 6817 genes in the dataset) it is possible to build an almost perfect classifier (Niranjan, personal communication). Building classifiers that describe the original data well, but perform badly on new data, is known as *overfitting*. Using simple classifiers instead of complex ones, unless there is a particular reason to use a particular complex one, may help to avoid overfitting.

In practice, classifiers are usually built based on the expression profiles of a relatively small set of informative genes. The success of the method relies to a large extent on how well this subset has been chosen. The performance also depends on how noisy the datasets are, whether the classes used to train the algorithm are correct, and whether the learning algorithm itself is appropriate for the particular data.

In the following subsections we discuss various types of classification algorithms and training methods.

4.4.2 Linear discriminants

Linear discriminants are one of the oldest classification methods and have been used for the past 30 years. We explain the basic idea in two-dimensional space, i.e. for the matrix

$$\begin{pmatrix} L_1 & L_2 & \cdots & L_m \\ x_{11} & x_{12} & \cdots & x_{1m} \\ x_{21} & x_{22} & \cdots & x_{2m} \end{pmatrix}$$

Remember that vectors $A_1 = (x_{11}, x_{21}), \ldots, A_m = (x_{1m}, x_{2m})$ can be represented as points in two-dimensional space. A *linear discriminant* in two dimensions is a straight line that separates (most) points labelled by 1 from those

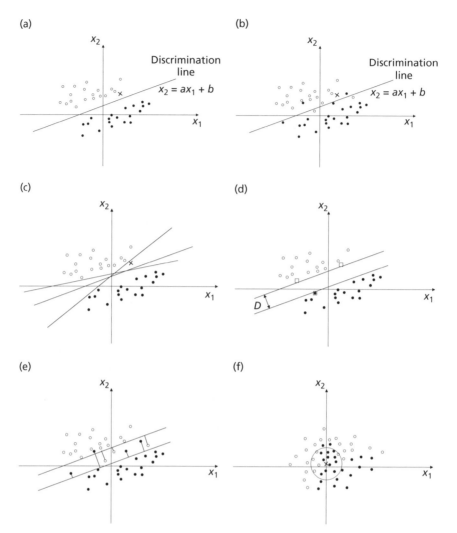

Figure 4.17 Linear discriminants. (a) The dots represent the expression profiles in two-dimensional space. Suppose the hollow dots are expression profiles labelled by 1, while the filled ones are labelled by −1. A straight line separates them perfectly. Given a new, unlabelled point, the method assigns the label −1 or 1 depending on whether it is placed below or above the regression line. For instance, the new point denoted by a cross will be assigned a label +1. (b) Note that in this case perfect separation by a straight line is not possible in two-dimensional space. (c) Many different linear discriminants can separate the filled and hollow dots in this case. Which one is the best in the sense that it is most likely to correctly classify a new point? Note that the cross is on different sides of different discriminants. (d) To solve the problem of the 'best' linear discriminant we can try to find the widest possible separation area between the points of different labels, i.e. to find maximal *D* such that the space of the width *D* separates all points correctly. This area is defined by only a few points, denoted here by squares, which are called support vectors. (e) If perfect separation is not possible, we can try to limit the total distance of points on the wrong side of the separation area, i.e. the total length of lines connecting them to the outer side of the separation area. All points on the wrong side of this area or in it become support vectors in this case. (f) A linear discriminant will not work here. The *K*-nearest neighbour method estimates the label from the labels of the *K* points nearest to the given point. For instance, if we take *K* = 16, then the label to the new point is denoted by the cross in the figure. This will be assigned a value −1 (i.e. filled) by majority voting.

labelled by −1 (Figure 4.17a,b). In some cases discrimination by a straight line perfectly classifies the data (Figure 4.17a), in some imperfectly (Figure 4.17b), in some not at all (Figure 4.17f).

In two dimensions a straight line is described by the equation $x_2 = ax_1 + b$. This means that all we have to do to build a linear classifier is to find the two parameters a and b. Note that a point $A_i = (x_{1i}, x_{2i})$ is situated above the discrimination line if $x_{2i} - ax_{1i} - b > 0$, and below the discrimination line if $x_{2i} - ax_{1i} - b < 0$. We would like to position the discrimination line to minimise the number of points $A_i = (x_{1i}, x_{2i})$, such that $L_i = -1$, but $x_{2i} - ax_{1i} - b > 0$, or $L_i = 1$, but $x_{2i} - ax_{1i} - b < 0$.

A common method used to find a discrimination line (i.e. to train a linear discriminant) is *least squares fit*, which is essentially the same method used in the linear regression-based normalisation algorithm described in the previous chapter. We define the residual sum of squares as

$$RSS(a, b) = \sum_{i=1}^{m} (L_i - (x_{2i} - ax_{1i} + b))^2$$

The least squares fit seeks to minimise $RSS(a, b)$, for the given set of labelled samples $(A_1, L_1), \ldots, (A_m, L_m)$, where $A_i = (x_{1i}, x_{2i})$. After we have found the regression line, given a new point $A_k = (x_{1k}, x_{2k})$ the classifier will assign to it label −1 if the point is below the regression line or +1 if it is placed above the line.

In three-dimensional space linear separation is accomplished by a plane, which can be described by three parameters, while in n dimensions the discrimination is accomplished by an $n - 1$-dimensional *hyperplane*, described by n parameters. In this way, the classification problem using linear discriminants boils down to finding a set of parameters describing a hyperplane separating points labelled by 1 from points labelled by −1 in n-dimensional space. The least squares fit can be generalised to the arbitrary number of dimensions in a relatively straightforward manner. Most standard statistical analysis software packages include this method.

4.4.3 Support vector machines

Consider the example in Figure 4.17c. Note that various different linear discriminants separate the filled and hollow dots perfectly. So how do we determine which describes the best separation of the data? One possible approach would be to use an area of separation instead of a line and to try to maximise the width of this area, as shown in Figure 4.17d. Note that the widest possible area of separation (i.e. the maximum D in Figure 4.17c) is in fact determined by only a few points (three points denoted by squares in the figure). These points are called *support vectors*.

Unfortunately it is not always possible to separate all the points using a linear discriminant (e.g. Figure 4.17b). In this case one possibility is to permit misclassified points, and, for instance, to limit the total permitted distance between misclassified points and the far side of the separation area. In this case, all points inside and on the 'wrong' side of the separation area are support vectors.

An alternative method would be to allow higher order surfaces instead of hyperplanes to separate the data (for instance in two dimensions one could use second order lines, i.e. parabolic lines). In fact, it turns out that instead of allowing for higher order discriminants directly, it is advantageous to project the original vector space to a higher dimensional space, and build the linear discriminant in the new space. This technique is known as *support vector machines* (SVMs).

In the SVM method each vector $\mathbf{X}$ in the original multidimensional space is transformed to a vector $\mathbf{h}(\mathbf{X})$ in a higher dimensional space. Hyperplanes in the higher dimensional space correspond to higher order surfaces in the original space. To use this method we need to define the function $\mathbf{h}$. It turns out that all that is needed to apply the SVM method is the definition of the dot product in the higher dimensional space for vectors, given in the original space (recall from Section 4.1 that for vectors $\mathbf{X} = (x_1, \ldots, x_n)$ and $\mathbf{Y} = (y_1, \ldots, y_n)$ the dot product is defined as $\mathbf{X} \cdot \mathbf{Y} = x_1 y_1 + \ldots + x_n y_n$). The function defining the dot product in the higher dimensional space is called the *kernel function*. One of the most often used kernel functions is $\mathbf{K}(\mathbf{X}, \mathbf{Y}) = (\mathbf{X} \cdot \mathbf{Y} + 1)$.

We will not discuss SVMs in any detail here (those interested in finding out more about the techniques may wish to consult Hastie *et al.*, 2001). However, it should be noted that SVMs have been widely used for supervised gene expression data analysis, both for sample and for gene classification. It has often been noted that SVMs give better classification performance than most other classification methods. Nevertheless, the success with which classification accuracy is achieved needs to be evaluated in relation to the complexity of the classifier.

SVMs have been used for yeast gene classification by Brown *et al.* (2000). Functional categories for some classes of genes, such as those encoding ribosomal proteins and histones (as labelled in the MIPS database; http://mips.gsf.de/proj/yeast/CYGD/db/index.html), were found to be readily predictable, while some other functional classes, such as kinases, are not predictable from gene expression data. SVMs showed better classification performance than all the other classification methods used. SVMs have been reported to show the best classification performance also for sample classification; however, the applications are usually based on independently selecting a few genes (e.g. 50 genes) that are used later.

4.4.4 *K*-nearest neighbour method

The linear discriminant method assumes that the labelled points can be separated by a hyperplane. In the example shown in Figure 4.17a,b this may be a reasonable assumption – only a few points are misclassified. However, in some cases this assumption is inappropriate; for instance, in the example shown in Figure 4.17f any linear discriminant will perform rather badly. A parabolic separator would perform better, thus SVMs may be more appropriate. A simple alternative to SVMs in some cases is the *K-nearest neighbour method*.

In the *K*-nearest neighbour method, the label of a new *a priori* unlabelled point is estimated from the labels of the closest K points (using a defined distance measure, e.g. Euclidean distance). Sometimes K is interpreted not as the number of points but as the radius around the given point. The simplest way to assign the label to a new point is so-called *majority voting*, where we simply take the K nearest points and count how many of them are labelled each way. *Weighted majority voting* assigns different weights to different dimensions. In this situation, the task of training the algorithm is to find the optimal weight for each dimension.

As discussed above for sample classification, there are many more gene expression profiles than samples to be classified. Moreover, most of the genes are likely to be irrelevant for the particular classification, and should be attributed weights equal to zero in the weighted majority voting. Unfortunately we do not know *a priori* which genes will be important for classification.

A heuristic method for selecting the genes that are likely to be important for analysis right at the beginning was proposed by Golub *et al.* (1999). They treated the class vector (see Section 4.4.1) as a gene expression vector characterising the sample, and looked for genes with expression profiles within distance K from the class vector. This effectively means looking for genes that are predominantly highly expressed in samples of one class, and expressed at low levels in the samples of the other class, and vice versa.

The distance can be measured as the correlation of the gene to the class vector. Let us assume that the given gene has a mean expression value within the samples of the first class equal to μ_1 and a standard deviation equal to σ_1. Let the mean expression value within the samples of the second class equal μ_2 and the variance σ_2. The correlation between the class vector and the given expression profile can be defined as the ratio $(\mu_1 - \mu_2)/(\sigma_1 - \sigma_2)$.

Next, one has to find whether the genes that correlate with the samples (in the sense described above) exist, and if they do, what is the optimal radius K. The first question can be addressed by finding whether the number of genes within the given distance K from the class vector is larger than expected by chance (which can be determined by comparing this number with that in randomised expression space). The second question can be addressed by trying out different values of K. It should be noted, however, that even if a gene perfectly correlates with the class vector, it will not necessarily prove a good classifier, since such a one gene-based classifier may be rather sensitive to possible noise in the data.

Once the K has been chosen (i.e. the genes that have higher than zero value in voting), the individual weights for the voting schema should be assigned to these genes. Heuristics weighting the genes by how well the gene correlates with the class vector was used in the above mentioned leukaemia classification example (Golub *et al.*, 1999).

4.4.5 Neural networks, decision trees and applications of classification

Neural networks and decision trees are two of the more common alternative types of classifiers. Neural networks describe a wide range of different classification methods, most of which are special cases of SVMs for appropriate kernel functions. We will not discuss these in detail here.

Decision trees are a simple classification method based on making the classification decision by asking a series of yes/no questions, where each successive question depends on the answer to the previous question. It has been demonstrated that decision trees can be used to predict functional classes of genes for yeast from gene expression profiles (Brazma, unpublished results). Yet another classification algorithm, based on selecting pairs of genes, has been proposed by Bo and Jonassen (2002).

Diagnostics is an obvious application for classification. If a classifier is based on a simple set of rules, such as comparing the expression levels of a relatively small set of genes with some threshold values, knowing these rules can aid in understanding the underlying mechanisms of the particular phenomena. For instance, having a list of genes whose expression levels are important for cancer classification may provide a clue to the mechanisms associated with particular cancer types. On the other hand, gene classification may also help to reveal which experimental conditions are related to various gene functions.

4.4.6 Partially supervised analysis

In partially supervised analysis we use the external information (i.e. the class vector) in the analysis as a guide, without necessarily trying to fit the data to the class vector perfectly. Two ways to achieve this are (i) including the class vector in the gene expression matrix as an additional dimension (as described in Section 4.4.1) and applying a clustering algorithm; and (ii) using a hierarchical clustering algorithm to build a dendrogram from the data, initially ignoring the class vector, and then looking for the clusters in the data best fitting the class vector.

Ben-Dor *et al.* (2000) used the second approach for classification of gene expression data from colon and ovarian cancer. After applying clustering to find the hierarchical structure in the data (i.e. the dendrogram), they used supervised learning to find the best thresholds at which to cut each subtree in the dendrogram to obtain the best correspondence between the clustered data and the class vector.

If we include the class vector in the initial analysis, the data analysis method can be changed from pure clustering to supervised analysis by varying the weight of the respective dimension in the distance measure. By making this weight equal to zero we have a completely unsupervised method, and by increasing it, and thus increasing the importance of the class vector in determining the clusters, we achieve supervised learning. This approach has been proposed in combination with the gene shaving algorithm in Hastie *et al.* (2000).

4.4.7 Class discovery

In sample classification we assume that the classes, such as acute myeloid leukaemias and acute lymphoblastic leukaemias, are defined *a priori*. In reality, however, samples that look morphologically the same in fact may represent several such groups each characterised by different expression profiles. Therefore subgroups may be revealed within a group of morphologically similar samples by gene expression analysis. Clustering can be used to try to discover such new classes. An example of such an approach has been used by Alizadeh *et al.* (2000), in which a heuristic clustering algorithm was applied to classify gene expression profiles. This permitted correct prediction of prognostic outcomes, supporting the hypothesis that the discovered clusters correspond to 'real' subclasses of tumours.

Application of the clustering approach to class discovery was used by Bittner *et al.* (Bittner *et al.*, 2000; Dougherty *et al.*, 2002), where subclasses of malignant melanoma were discovered. A different approach to class discovery has been described by Heydebreck *et al.* (2001).

4.5 Time series analysis

Time series experiments provide a particular type of gene expression profile, revealing information about the order and the time scale of the expression events. There are a number of ways one can treat time series that would not be meaningful for other types of expression profile. An obvious example of time series analysis is directed towards finding periodicity or a trend (Figure 4.18a). This approach was used by Spellman *et al.* (1998) for identifying genes whose expression correlates with known events in the cell cycle of yeast (so-called cell cycle regulated genes).

For an expression level of a particular gene the *period T* can be defined as the time after which the expression level of the gene repeats. The *frequency* is defined as the inverse of the period, i.e. $f = 1/T$. The *amplitude* of the fluctuation is the difference between the maximum and minimum values. If the average expression value systematically increases or decreases, we can say that the time course has a *trend*.

Given a time series we may be interested to find out whether it is periodic. A common method used in signal processing for finding periodic components in data is Fourier analysis. Time series in microarray experiments typically do not contain enough time points to make application of this powerful method necessary – for a limited set of data points we can simply look for correlations between the original time course and time courses obtained from the original by shifting it by 1, 2, . . . time points. If we obtain a strong correlation upon shifting by k time points, this means that there is a period equal to k.

To be able to detect the true fluctuation period we have to sample the time points with a frequency at least as high as that of the real fluctuation. For instance, if we perform the measurements with a frequency that equals half of the true frequency, we would conclude from the measurements that the true frequency is half of what it actually is (e.g. Figure 4.18b). A sampling frequency that is close to but slightly larger than the true one may also lead to the 'detection' of a false trend.

In microarray experiments we measure expression levels of thousands of

(a)

Expression level

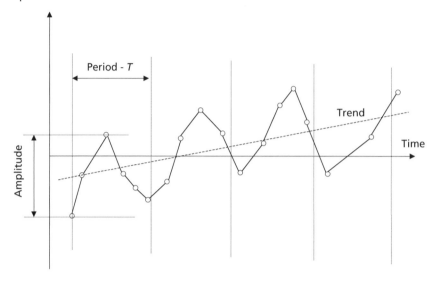

(b)

Expression level

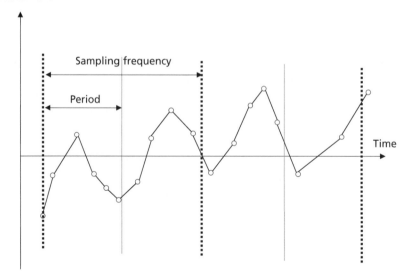

Figure 4.18 Time series analysis. (a) Time course: periodicity and trend. If the fluctuation period equals T, then the fluctuation frequency is defined as $f = 1/T$. (b) Time course: if the gene expression fluctuation frequency is larger than the sampling frequency (i.e. the gene expression fluctuation period is smaller than the sampling period) we will not be able to identify the correct period. (c) Two time series fluctuations with the same frequency may be shifted in phase. (d) Time warping (see text for explanation).

(c)

Expression level

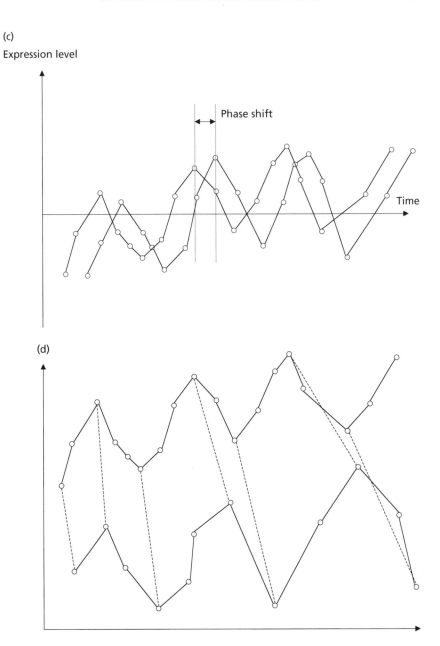

Figure 4.18 *Continued*

genes in parallel. Looking for periodicity in time series of each of the thousands of genes separately, we are likely to find periodic behaviour for some genes which is simply the result of noise in the data. Luckily, it is unlikely that in any biological system many genes have different periods – in most cases we are looking for periods that are common to many genes, as the possible result of co-regulation. Moreover, often such periods will also correlate with some morphological changes in the system, such as the cell cycle. In the above mentioned paper by Spellman *et al.*, about 800 genes that change with the cell cycle were detected.

Genes changing with the same period can still have different fluctuation *phases* (Figure 4.18c). Clustering algorithms will usually cluster together genes that have similar periods and phases. For instance, in the cell cycle experiment, five major 'phase groups' associated with G1, S, G2, M and M/G1 phases can be distinguished (the exact distinction is somewhat arbitrary).

Sometimes we may wish to compare gene expression time series from different experiments corresponding to similar biological situations. An example of such an approach is comparing gene expression during the cell cycle for cell cultures synchronised using different methods. In this case we cannot be sure that the periods will be exactly the same. If some of the genes involved in the process under study are known, we can 'synchronise' the periods, by comparing the expression levels of these known genes (for instance by relating each successive maximum and minimum in both series). However, in general we may need to use a distance measure, for which the time points from both time series that correspond to each other are not defined *a priori*. This is known as *time warping* (Figure 4.18d).

Time warping has been studied in different contexts, and various algorithms – usually based on so-called dynamic programming – have been developed. For instance, one such algorithm by Krusksal and Liberman (1999) has been applied to gene expression data analysis (Aach and Church, 2001). In this study, series from yeast cell cycle samples synchronised by different methods were compared.

4.6 Visualisation

Visualisation is a powerful data mining technique for finding patterns in data, and has been used extensively for gene expression data analysis. For instance, a simple technique called a Venn diagram (Figure 4.9b) can be used to demonstrate the extent that clusters of gene expression data overlap as well as to show the overlaps of gene expression data clusters with GO functional annotation.

Owing to the high dimensionality of gene expression matrices, visualisation is usually coupled with techniques allowing reduction of the dimensionality in the data, such as clustering or principal component analysis. The most popular visualisation techniques are heat maps, first introduced for gene expression data analysis by Michael Eisen (Eisen *et al.*, 1998). Essentially a heat map is simply a representation of the gene expression matrix using colour coding. For instance, for log ratio matrices, the positive values can be represented using the colour red, negative values using the colour green. The intensity of the colour represents the absolute value (i.e. 0 is represented by black), as shown in Plate 4.4 (facing p. 88). Heat maps are typically used in association with clustering. If hierarchical clustering is used the dendrogram can be given using the same visualisation. Various software packages provide a means for zooming-in, thus gradually revealing increasingly more detailed information, such as gene names and functional annotation.

A different popular way to depict gene expression profiles and clusters of profiles is through the use of *profile graphs* (see Plate 4.5, facing p. 88). Profile graphs can be obtained by plotting expression values on the vertical axis, samples on the horizontal axis, and joining the points corresponding to the same genes in different samples. Note that profile graphs do not necessarily represent time series. If they are used to represent other types of gene expression data, the lines joining sample expression values should not be interpreted as interpolations of expression values, but only as a means to relate the same genes in different samples.

A rather different visualisation method makes use of the reduction in the dimensionality achieved by applying principal component analysis (as discussed in Section 4.2.4). If most of the variability in the data can be attributed to two or three principal axes, these can be used for visualising data in two or three dimensions. Colour coding can be used to plot the density of the genes in the reduced space to visualise clusters. As an example, such a density plot in two principal dimensions is shown in Plate 4.6 (facing p. 88), where two clusters can be identified.

Another way in which covariance or gene expression datasets has been represented is using a topo, or *gene expression terrain map* (see Plate 4.7, facing p. 88). The expression levels of genes are used to calculate the covariance between datasets in large numbers of experiments. The covariance is represented in two dimensions, such that closely related data are together and the altitude of the 'gene mountain' represents the density of genes at that site.

Correspondence analysis is a method that uses principal component analysis in a chi-square distance matrix and visualises the two or three principal axes of gene and sample space in the same diagram. This allows us to assess which experimental conditions are most important for defining which groups of genes, and vice versa (i.e. which genes are most important in which of the experimental conditions). Cell cycle data from Spellman *et al.* (1998) visualised in this way are shown in Plate 4.8 (facing p. 88). For more details on correspondence analysis see Fellenberg *et al.* (2001).

A rather different visualisation method is based on depicting the relationships among genes in the form of networks (see Section 4.2.6). Graph layout algorithms can be used for visualising various gene expression networks, like that shown in Figure 4.8.

Visualisation methods can also be used to combine gene expression data with other relevant data. We have already seen how heat maps can be used in combination with GO annotation (Plate 4.2, facing p. 88). Another simple visualisation technique, but of a different kind, is *grid display*. This is often used to display gene expression levels in relation to the position of the gene on the array, allowing gross array defects and local variation to be detected easily (Figure 4.19a). *Chromosome displays* can be used to visualise the expression of genes in relation to their position along the chromosome. This is useful for identifying whether gene expression correlates with chromosomal location (Figure 4.19b).

Visualisation can also be used to find if sets of co-expressed genes contain particular sequence elements in their promoters. This method has been popular in the analysis of transcriptional regulation in yeast, since it is believed that promoter sequences are typically within a few hundred base pairs upstream of the genes they regulate – see Plate 4.9 (facing p. 88) and Plate 4.10 (facing p. 88) for sequence patterns found in yeast.

Similarly, visualisation is used to highlight the expressed genes in metabolic pathways; an example is shown in Figure 4.20.

Visualisation can also be applied to distance matrices. Plate 4.1 (facing p. 88) shows a coloured version of a covariance matrix in which the Pearson correlation obtained from a comparison of two genome-wide datasets is represented. The datasets that affect similar sets of genes, and are therefore closely related, can be seen as blocks of pink squares.

4.7 Downstream from expression profile analysis

The types of supervised, unsupervised and data visualisation methods described above are only the first steps in expression data analysis. First, co-expressed genes are of interest because there is evidence that many functionally related genes are co-expressed. This has been widely demonstrated for genes involved in processes such as translation and elongation and for subunits of multi-subunit complexes, such as ribosomal protein subunits and histones (Brown and Botstein, 1999). Second, co-expression may be the result of co-regulation; e.g. two genes that are co-expressed may be regulated by a common transcriptional regulator. This can be used to study gene regulation mechanisms, as described below.

Comparisons between groups of genes, such as those identified by clustering, can be performed in a number of ways, e.g. comparison with other groups of co-expressed genes, or comparison with annotated characteristics of genes or the literature. The object of the exercise is usually to see if there are correlations between the groups under comparison that are greater than might be expected by chance. In some instances, the genes in a cluster may fall into subgroups.

(a)

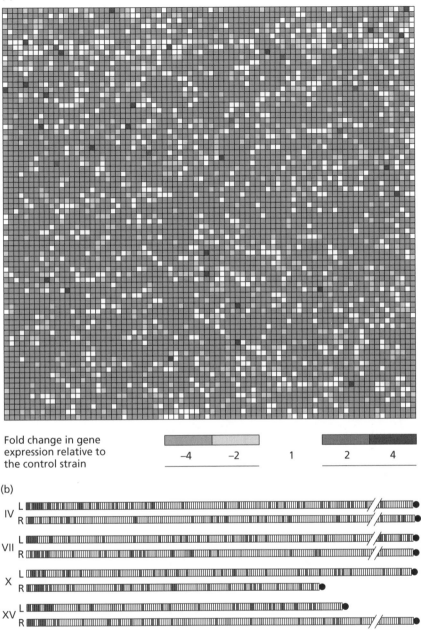

Fold change in gene
expression relative to
the control strain

−4   −2   1   2   4

(b)

IV L
    R

VII L
    R

X   L
    R

XV L
    R

Figure 4.19 Examples of grid and chromosome displays. Both representations of the data show
the results of a comparison between a mutant yeast strain and its isogenic wild type counterpart. In
a grid display, each of the boxes represents the relative expression level of a gene. The shade of the
box indicates whether there has been an increase in the amount of mRNA encoding the gene or a
decrease in the mRNA relative to the control strain. Panel (a) shows the effect of inactivating RNA
polymerase II, the enzyme responsible for the transcription of most protein-encoding genes in
yeast, using a grid display. The transcript levels appear to drop markedly for most genes because
transcription of these messages ceases upon activation of the enzyme. Panel (b) shows the effect of
inactivating histone H4 using a chromosome display. Here the genes are ordered according to
their normal positions along each chromosome in half-chromosome segments, beginning at the
telomere of the chromosome (left or right) and ending at the centromere (black circle). ((a)
reprinted from *Cell* (Holstege *et al.*), Copyright 1998, with permission from Elsevier Science. (b)
reprinted with permission from *Nature* (Wyrick *et al.*), Copyright 1998 Macmillan Magazines
Ltd.)

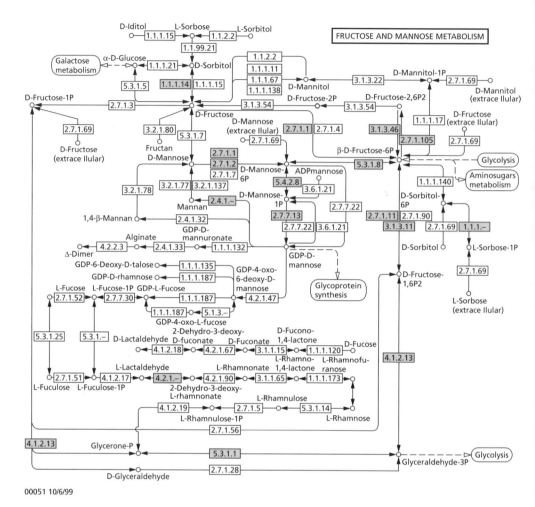

Figure 4.20 Metabolic pathway from the KEGG database (http://www.genome.ad.jp/kegg/), with genes belonging to a particular cluster of expression profiles highlighted.

00051 10/6/99

4.7.1 Identification of regulatory signals

It seems reasonable to hypothesise that genes with similar expression profiles, i.e. genes that are co-expressed, may share common regulatory mechanisms, i.e. may be co-regulated, and this can be applied to finding groups of potentially co-regulated genes and to search for putative regulatory signals. The outline of such a discovery method is as follows: (i) cluster the genes based on a selection of expression measurements; (ii) extract putative promoter sequences for the genes in the clusters; (iii) search for sequence patterns statistically overrepresented in these clusters; and (iv) assess the quality of discovered patterns using some statistical significance criteria.

One of the most difficult steps in this approach is identifying the putative pro-moter sequences. In higher eukaryotes gene regulatory sequences can be found far from the coding regions of the gene, and are difficult to detect. For yeast, however, there is evidence that most regulatory regions are relatively close to the coding regions, typically within about 600–800 base pairs upstream from the translation start point. Therefore upstream sequences within these limits can be used as putative promoter regions.

A systematic application of this approach has been reported for the yeast *Saccharomyces cerevisiae* using a public dataset from Stanford University (Eisen *et al.*, 1998), combining various yeast expression experiments (80 conditions for 6221 genes). In the absence of theoretically 'correct' similarity measures and clustering algorithms, the simplest measure was selected and different clus-terings carried out. All genes were clustered based on their expression profiles by the *K*-means clustering algorithm using Euclidean distance. In total over 900 separate clusterings were obtained, and clusters containing between 20 and 100 genes were selected, totalling over 52,100 different (mostly overlap-ping) clusters (Vilo *et al.* 2000).

For each cluster of genes the set of upstream sequences of length 600 base pairs was selected, and all sequence patterns that could be represented as sub-strings of any length occurring in at least 10 sequences in a cluster were scored according to the binomial probability of their occurrences in the cluster (the background probability was estimated based on the number of occurrences of each pattern in upstream sequences of all 6221 genes). To determine the statis-tical significance threshold for the patterns, the last step was repeated on ran-domised data. After removal of patterns common to highly homologous upstream sequences, 1498 significant patterns remained. These patterns can be matched back to the gene upstream sequences and visualised in conjunction with gene expression data, and some of them are shown in Plate 4.10 (facing p. 88) and Figure 4.20.

As 1498 patterns is still too many for each to be studied individually, they were clustered using a similarity measure based on common information con-tent. This produced 62 clusters of similar patterns. For each cluster of patterns an approximate alignment and a consensus pattern was calculated.

The patterns identified were evaluated against known transcription factor binding sites. All 1498 'interesting' patterns were matched against experi-mentally verified DNA binding sites of yeast as given in SCPD databases (http://cgsigma.cshl.org/jian/).

Of the 62 clusters of patterns, 48 had matches in SCPD and 14 did not have a match at any site reported in the SCPD database. Table 4.9 shows the partial consensus patterns that have been calculated from pattern alignments for these 14 clusters. The fact that 48 out of 62 pattern classes match experimentally verified yeast transcription factor binding sites indicates the validity of the described computational discovery method. However, the most interesting patterns are likely to be the ones that do not have matches in the known binding sites. Automatic or semiautomatic generation of such hypotheses is one of the main tasks of bioinformatics and data mining approaches.

Table 4.9 Consensi of the pattern clusters that do not have matches in the SCPD database (see text). Brackets mean that any of the nucleotides inside can occur in the pattern (e.g. [AT] means that either A or T can occur at that position and have been observed in data with at least 25% frequency). Inside the groups nucleotides are ordered based on their frequency in the data. Lowercase is used when the majority of the patterns do not have any nucleotide at that position, i.e. when the most frequent nucleotide in the respective column is a dash.

---

aaTCTTCATGt
cgTACCTCTa
gACAGCTAc
tAT[TAC]GTTAAgc
ACTTTATTT
[ag]TAACTT[AT]Ca
TATCGAG (singleton)
t[ta]CGAATA[AG]aaaa
[ta]TGCATGAAc
a[TG][GC]GTATAc
[ag][ga][AG]ATATG[TG][ga][ag]g
tag[AG]TAGA[TA]A[ga]aaaa
ATCCAAGAg
tTTTTCTG[CT][TA]c

References

Aach, J., and Church, G. M. (2001). Aligning gene expression time series with time warping algorithms. *Bioinformatics* **17**, 495–508.

Alizadeh, A. A., Eisen, M. B., Davis, R. E., Ma, C., Lossos, I. S., Rosenwald, A., Boldrick, J. C., Sabet, H., Tran, T., Yu, X., *et al.* (2000). Distinct types of diffuse large B-cell lymphoma identified by gene expression profiling. *Nature* **403**, 503–511.

Alon, U., Barkai, N., Notterman, D. A., Gish, K., Ybarra, S., Mack, D., and Levine, A. J. (1999). Broad patterns of gene expression revealed by clustering analysis of tumor and colon tissues. *Proceedings of the National Academy of Sciences of the United States of America* **96**, 6745–6750.

Alter, O., Brown, P. O., and Botstein, D. (2000). Singular value decomposition for genome-wide expression data processing and modeling. *Proceedings of the National Academy of Sciences of the United States of America* **97**, 10101–10106.

Ben-Dor, A., Bruhn, L., Friedman, N., Nachman, I., Schummer, M., and Yakhini, Z. (2000). Tissue classification with gene expression profiles. In *Fourth Annual International Conference on Computational Molecular Biology RECOMB-2000*, R. Shamir, S. Miyano, S. Istrail, P. Pevzner, and M. Waterman, eds., pp. 54–64 (ACM Press, Tokyo).

Bittner, M., Meltzer, P., Chen, Y., Jiang, Y., Seftor, E., Hendrix, M., Radmacher, M., Simon, R., Yakhini, Z., Ben-Dor, A., *et al.* (2000). Molecular classification of cutaneous malignant melanoma by gene expression profiling. *Nature* **406**, 536–540.

Bo, T. H., and Jonassen, I. (2002). New feature subset selection procedures for classification of expression profiles. *Genome Biology* **3**, 0017.1–0017.11.

Brazma, A., Jonassen, I., Vilo, J., and Ukkonen, E. (1998). Predicting gene regulation elements in silico on a genomic scale. *Genome Research* **8**, 1202–1215.

Brown, M. P. S., Grundy, W. N., Lin, D., Cristianini, N., Sugnet, C. W., Furey, T. S., Ares, M. J., and Haussler, D. (2000). Knowledge-based analysis of microarray gene expression data by using support vector machines. *Proceedings of the National Academy of Sciences of the United States of America* **97**, 262–267.

Brown, P. O., and Botstein, D. (1999). Exploring the new world of the genome with DNA microarrays. *Nature Genetics (Supplement)* **21**, 33–37.

DeRisi, J. L., Iyer, V. R., and Brown, P. O. (1997). Exploring the metabolic and genetic control of gene expression on a genomic scale. *Science* **278**, 680–686.

Dopazo, J., and Carazo, J. M. (1997). Phylogenetic reconstruction using a growing neural network that adopts the topology of a phylogenetic tree. *Journal of Molecular Evolution* **44**, 226–233.

Dougherty, E. R., Barrera, J., Brun, M., Kim, S., Cesar, R. M., Chen, Y., and Bittner, M. (2002). Inference from clustering with application to gene-expression. *Journal of Computational Biology* **9**, 105–126.

Efron, B., and Tibshirani, R. (1991). Statistical data analysis in the computer age. *Science* **253**, 390–395.

Eisen, M. B., Spellman, P., Botstein, D., and Brown, P. O. (1998). Cluster analysis and display of genome-wide expression patterns. *Proceedings of the National Academy of Sciences of the United States of America* **95**, 14863–14868.

Everitt, B. S., and Dunn, G. (2001). *Applied Multivariate Data Analysis*, 2nd edn. (Edward Arnold, London).

Fellenberg, K., Hauser, N. C., Brors, B., Neutzner, A., Hoheisel, J. D., and Vingron, M. (2001). Correspondence analysis applied to microarray data. *Proceedings of the National Academy of Sciences of the United States of America* **98**, 10781–10786.

Filkov, V., Skiena, S., and Zhi, J. (2001). Analysis techniques for microarray time-series data. Paper presented at *Fifth Annual Conference on Research in Computational Molecular Biology* (ACM Press, New York).

Francis, A. (1988). *Advanced Level Statistics: An Integrated Course*, 2nd edn (Stanley Thorne, UK).

GO consortium (Ashburner, M., Ball, C. A., Blake, J. A., Botstein, D., Butler, H., Cherry, M. J., Davis, P. A., Dolinski, A. P., Dwight, S. S., *et al.*) (2000). Gene Ontology: tool for the unification of biology. *Nature Genetics* **25**, 25–29.

Golub, T. R., Slonim, D., Tamayo, P., Huard, C., Gaasenbeek, M., Mesirov, J. P., Coller, H., Loh, M. L., Downing, J. R., Caligiuri, M. A., *et al.* (1999). Molecular classification of cancer: class discovery and class prediction by gene expression monitoring. *Science* **286**, 531–537.

Harary, F. (1969). *Graph Theory* (Addison-Wesley, Reading, MA).

Hartigan, J. A. (1975). *Clustering Algorithms* (John Wiley & Sons, New York).

Hastie, T., Tibshirani, R., Eisen, M., Alizadeh, A., Levy, R., Staudt, L. M., Chan, W. C., Botstein, D., and Brown, P. O. (2000). 'Gene shaving' as a method for identifying distinct sets of genes with similar expression patterns. *Genome Biology* **1**, 0003.0001–0003.0021.

Hastie, T., Tibshirani, R., and Friedman, J. H. (2001). *The Elements of Statistical Learning: Data Mining, Inference, and Prediction* (Springer Series in Statistics, Springer-Verlag, New York).

Herrero, J., Valencia, A., and Dopazo, J. (2001). A hierarchical unsupervised growing neural network for clustering gene expression patterns. *Bioinformatics* **17**, 126–136.

Heydebreck, A., Huber, W., Poustka, A., and Vingron, M. (2001). Identifying splits with clear separation: a new class discovery method for gene expression data. *Bioinformatics* **17**, s107–s114.

Holstege, F. C., Jennings, E. G., Wyrick, J. J., Lee, T. I., Hengartner, C. J., Green, M.

R., Golub, T. R., Lander, E. S., and Young, R. A. (1998). Dissecting the regulatory circuitry of a eukaryotic genome. *Cell* **95**, 717–728.

Jain, A. K., Murty, M. N., and Flynn, P. J. (1999). *ACM Computing Surveys*, Vol. 31 (ACM Press, New York).

Kim, S. K., Lund, J., Kiraly, M., Duke, K., Jiang, M., Stuart, J. M., Eizinger, A., Wylie, B. N., and Davidson, G. S. (2001). A gene expression map for *Caenorhabditis elegans*. **293**, 2087–2092.

Kohonen, T. (1990). The self-organizing map. *Proceedings of the IEEE* **9**, 1464–1479.

Krusksal, J. N., and Liberman, M. (1999). The symmetric time-warping problem: from continuous to discrete. In *Time Warps, String Edits, and Macromolecules: The Theory and Practice of Sequence Comparisons*, D. Sankoff and J. N. Kruksal, eds., pp. 125–161 (C.S.L./I.; reissued 2000, University of Chicago Press).

Lance, G. N., and Williams, W. T. (1967). Mixed-data classificatory programs. II. Divisive systems. *Australian Computer Journal* **1**, 82–85.

Lee, T. I., Causton, H. C., Holstege, F. C. P., Shen, W.-C., Hannett, N., Jennings, E. G., Winston, F., Green, M. R., and Young, R. A. (2000). Redundant roles for the TFIID and SAGA complexes in global transcription. *Nature* **405**, 701–704.

Legendre, P., and Legendre, L. (1998). *Numerical Ecology*, 2nd edn. (Elsevier, Amsterdam).

Li, M., and Vitanyi, P. (1993). *An Introduction to Kolmogorov Complexity and its Applications* (Springer-Verlag, New York).

Mangiameli, P., Chen, S. K., and West, D. (1996). A comparison of SOM neural network and hierarchical clustering methods. *European Journal of Operational Research* **93**, 402–417.

Rousseeuw, P. J. (1987). Silhouettes: A graphical aid to the interpretation and validation of cluster analysis. *Journal of Computational and Applied Mathematics* **20**, 53–65.

Rung, J., Schlitt, T., Brazma, A., Freivalds, K., and Vilo, J. (2002). Building and analysing genome-wide gene disruption networks. *Bioinformatics* **18**, S202–S210.

Sharan, R., and Shamir, R. (2000). CLICK: A clustering algorithm with applications to gene expression data. Paper presented at *Eighth International Conference on Intelligent Systems for Molecular Biology* (American Association for Artificial Intelligence, Melno Park, CA).

Spellman, P. T., Sherlock, G., Zhang, M. Q., Iyer, V. R., Anders, K., Eisen, M. B., Brown, P. O., Botstein, D., and Futcher, B. (1998). Comprehensive identification of cell-cycle regulated genes of the yeast *Saccharomyces cerevisiae* by microarray hybridization. *Molecular Biology of the Cell* **9**, 3273–3297.

Strang, G. (1993). *Introduction to Linear Algebra*, 2nd edn. (Wellesley, Cambridge).

Tamayo, P., Slonim, D., Mesirov, J., Zhu, Q., Kitareewan, S., Dmitrovsky, E., Lander, E. S., and Golub, T. R. (1999). Interpreting patterns of gene expression with self-organizing maps: methods and application to hematopoetic differentiation. *Proceedings of the National Academy of Sciences of the United States of America* **96**, 2907–2912.

Tavazoie, S., Hughes, D., Campbell, M. J., Cho, R. J., and Church, G. M. (1999). Systematic determination of genetic network architecture. *Nature Genetics* **22**, 281–285.

Tibshirani, R., Hastie, T., Eisen, M. B., Ross, D. T., Botstein, D., and Brown, P. O. (1999). Clustering methods for the analysis of DNA microarray data. In *Technical Report* (Stanford University, Department of Statistics, Stanford, CA; http://www-stat.stanford.edu/~tibs/research.html).

Törönen, P., Kolehmainen, M., Wong, G., and Castrén, E. (1999). Analysis of gene expression data using self-organizing maps. *FEBS Letters* **451**, 142–146.

Troyanskaya, O., Cantor, M., Sherlock, G., Brown, P. O., Hastie, T., Tibshirani, R., Botstein, D., and Altman, R. B. (2001). Missing value estimation methods for DNA microarrays. *Bioinformatics* **17**, 520–525.

Tusher, V. G., Tibshirani, R., and Chu, G. (2001). Significance analysis of microarrays applied to the ionizing radiation response. *Proceedings of the National Academy of Sciences of the United States of America* **98**, 5116–5121.

Vilo, J., Brazma, A., Jonassen, I., Robinson, A., and Ukkonen, E. (2000). Mining for putative regulatory elements in the yeast genome using gene expression data. Paper presented at *Eighth International Conference on Intelligent Systems for Molecular Biology* (American Association for Artificial Intelligence, Melno Park, CA).

Wyrick, J. J., Holstege, F. C. P., Jennings, E. G., Causton, H. C., Shore, D., Grunstein, M., Lander, E. S., and Young, R. A. (1998). Chromosomal landscape of nucleosome-dependent gene expression and silencing in yeast. *Nature* **402**, 418–421.

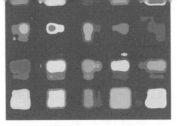

Appendix: non-commercial software

Many tools for the analysis of gene expression data have been developed and are free for academic use. Despite their utility and a general need for additional data analysis tools, many of these programs are not widely used (Cluster, XCluster and Treeview, developed at Stanford University, and D-Chip are notable exceptions). There are, however, several excellent programs that extend the range of available tools and, in some cases, offer functionality not found in commercial data analysis packages. A few of these programs are described in this Appendix, along with information on how to obtain access to them (for clarity 'http://' has been omitted from the start of URLs). There are many sites that provide links to other software for data analysis: three that are expecially worth looking at are ep.ebi.ac.uk/Links.html, linkage.rockefeller.edu/wli/microarray/soft.html and www.bioconductor.org.

A.1 Statistical analysis

BCLUST

bioinformatics.med.yale.edu/
A program to assess reliability of gene clusters from expression data by using a consensus tree and bootstrap resampling method, as described by Zhang and Zhao (2000).

BRB ArrayTools

linus.nci.nih.gov/BRB-ArrayTools.html

BRB ArrayTools is an integrated package, developed by Richard Simon and Amy Peng at the National Cancer Institute, for the visualisation and statistical analysis of DNA microarray gene expression data. The package uses an Excel front end with integrated analytical and visualisation tools.

Cyber-T

genebox.ncgr.org/genex/cybert/

Cyber-T is a web interface designed to detect changes in gene expression in large scale gene expression experiments. It operates on a set of functions written in R by Tony Long and Harry Mangalam, and can be run either via the Web or via direct manipulations in R. Cyber-T employs statistical tests, based on the t-test, to identify statistically significant differences between sample sets. These t-tests employ either the observed variance among replicates within treatments or a Bayesian estimate of the variance among replicates within treatments.

GEDA (Gene Expression Data Analysis)

www.biostat.wisc.edu/geda/eba.html

The GEDA algorithm uses an empirical Bayesian method to estimate true differential expression and identify significant differential expression in data generated using spotted arrays (Newton *et al.*, 2001). Estimates of gene expression changes are derived in a model that accounts for measurement error and fluctuations in absolute gene expression levels. GEDA is available in S-plus.

MA-ANOVA programs for microarray data

www.jax.org/research/churchill/software/anova/

MA-ANOVA is a set of functions for the analysis of variance of microarray data. The program is written in Matlab and includes functions for the calculation of log ratios, ratio–intensity plots, permutation tests and bootstrapping of confidence intervals.

PaGE (Patterns from Gene Expression)

www.cbil.upenn.edu/PaGE/

PaGE is a tool to attach descriptive and easily interpretable expression patterns to genes represented in microarray or macroarray data. The algorithm incorporates a novel method to identify differentially expressed genes between two sample types with attached measures of confidence (Manduchi *et al.*, 2000). The method is also described in Grant *et al.* (2002).

SMA (Statistics for Microarray Analysis)

stat.berkeley.edu/users/terry/zarray/Software/smacode.html

The SMA code includes simple R functions for within-slide normalisation for two-channel microarray data and identification of single differentially expressed genes. Further information is available in two technical reports (Dudoit *et al.*, 2000; Yang *et al.*, 2001).

VERA (Variability and Error Assessment) and SAM (Significance of Array Measurement)

www.systemsbiology.org/VERAandSAM/
VERA estimates the parameters of a statistical model that describes multiplicative and additive errors influencing a two-channel microarray experiment, using a maximum likelihood method. This is combined with the SAM program (www-stat.stanford.edu/~tibs/SAM/index.html; Tusher *et al.* 2001) to calculate the significance of fold change in gene expression (Ideker *et al.*, 2000).

A.2 Normalisation, clustering and classification

CLEAVER (Classification of Expression Arrays)

classify.stanford.edu/
Software developed at Stanford Biomedical Informatics for analysis of microarray data, including classification using discriminant analysis, *K*-means clustering and visualisation using principal component analysis. Calculations are carried out on the server side. Examples of input data files are available from smi-web.stanford.edu/projects/helix/pubs/pda/.

CLICK

CLICK and EXPANDER

www.cs.tau.ac.il/~rshamir/expander/expander.html
CLICK (CLuster Identification via Connectivity Kernels) and EXPANDER (EXPression Analysis and Display ManagER) is a java-based tool for clustering and visualising gene expression data. It has implementations of several clustering algorithms including K-means, self-organising maps, hierarchical clustering and CLICK. CLICK is a novel clustering algorithm, developed by Roded Sharan and Ron Shamir in which no prior assumptions are made on the structure or the number of the clusters. The algorithm uses techniques from statistics and graph theory to identify tight groups of highly similar elements (kernels), which are likely to belong to the same cluster. Heuristic procedures are subsequently used to expand the kernels.

CLUSFAVOR

mbcr.bcm.tmc.edu/genepi/
CLUSter and Factor Analysis using Varimax Orthogonal Rotation performs

cluster and factor analysis of gene expression data obtained from cDNA microarrays (Peterson, 2002). The user can perform cluster analysis and varimax orthogonal rotation, view dendrograms, and run factor analysis on selected cluster-specific genes. An optional output contains matrices for the input data, distance matrices, factor loadings, eigen-values, eigen-vectors, and the percentage of total variation for genes within a cluster.

Gene Cluster

rana.lbl.gov/EisenSoftware.htm and www.microarrays.org/software.html
Gene Cluster permits filtering, log transformation and mean centring of gene expression data, and downstream analysis including hierarchical clustering, self-organising maps and K-means clustering (Eisen *et al.*, 1998). A useful feature is that the data can be weighted, so samples for which there are replicates can be 'down-weighted' when calculating the distances between samples. The output is usually visualised using Treeview.

Gene Cluster and Treeview are frequently used together to generate characteristic red–green images. These typically represent changes in gene expression of clustered expression data (e.g. Plate 4.8, facing p. 88).

GeneCluster

www-genome.wi.mit.edu/cancer/software/software.html
GeneCluster is a program for generating self-organising maps. Data can be normalised and filtered within the program and the output provides a graphical representation of the clusters. Use of the algorithm is described in Tamayo *et al.* (1999). GeneCluster 2.06 extends these capabilities and includes supervised classification, gene selection and permutation test methods. It implements the methodology used in Golus *et al.* (1999).

Kimono (*K*-means Integrated Models for Oligonucleotide Arrays)

www.fruitfly.org/~ihh/kimono/
Kimono is a software package for finding regulatory elements in promoter sequences using quantitative expression data. The algorithm uses K-means-based clustering to group a dataset of promoter sequences and associated expression profiles.

Plaid

www-stat.stanford.edu/~owen/plaid/
Plaid implements a novel algorithm for clustering of gene expression data in which a gene can belong to more than one cluster and gene clusters may be defined with respect to samples. Further information can be obtained in a

technical report www.stat.stanford.edu/~owen/reports/ and in the accompanying paper (Lazzaroni and Owen, 2002).

RCluster

genex.ncgr.org/genex/rcluster/help.html
RCluster is a Web interface for a collection of clustering routines written in R for analysis of gene expression data. The algorithm permits hierarchical agglomerative clustering using a number of distance measures and linkage methods.

SOTA (Self-Organising Tree Algorithm)

bioinfo.cnio.es/sotarray/
The SOTA program uses a divisive hierarchical method for clustering data (Dopazo and Carazo, 1997). The advantage of this approach is that clustering can be stopped at a point appropriate for the individual dataset. This point may be chosen based on a variability threshold, or at a predefined number of clusters. SOTA runs on a server, and is described in Herrero *et al.* (2001).

SVDMAN (Singular Value Decomposition Microarray Analysis)

public.lanl.gov/mewall/svdman/
This algorithm permits singular value decomposition using two new methods. One is a threshold method for identifying groups of genes and the other measures confidence in SVD analysis (Wall *et al.*, 2001).

SNOMAD (Standardization and Normalization of Microarray Data)

pevsnerlab.kennedykrieger.org/snomad.htm
SNOMAD is a collection of Web-based tools for the standardisation and gene normalisation of expression data (Colantuoni *et al.*, 2002). The program includes visualisation and transformation tools, including those that correct for local bias and variance in gene expression measurements.

Xcluster

genome-www.stanford.edu/~sherlock/cluster.html
A program for filtering and clustering data. The program includes an option to partition the data into sets of related genes using self-organising maps or *K*-means, before carrying out hierarchical clustering. In addition, there is a 'node flipping' option, that permits rotation about the nodes of clustered data such that the two most similar outermost members of the nodes are placed adjacent to each other. The output can be visualised using Treeview and maxdView.

A.3 Visualisation

GenMAPP (Gene MicroArray Pathway Profiler)

www.genmapp.org/
GenMAPP is an application for the visualisation and display of gene expression data, using maps representing biological pathways and groupings of genes. The program can be used to display data from human, mouse, rat and yeast, and there are links to species-specific gene databases.

TreeArrange and Treeps

monod.uwaterloo.ca/software/
TreeArrange and Treeps are programs for reordering and displaying microarray data. Treeps is a tool for displaying expression array data and associated hierarchical clustering, while TreeArrange reorders the branches of the 'tree'/dendrogram so that similar 'leaves' are placed together. The algorithm is described in Biedl *et al.* (2001). The program can be run using the output files from Gene Cluster and Xcluster.

Treeview

rana.lbl.gov/EisenSoftware.htm
Treeview is a visualisation tool for the representation of gene expression data, typically $\log_2$ fold change in gene expression, using a gradient of red through black to green (Eisen *et al.*, 1998). Treeview is often used with Gene Cluster to display the results of cluster analysis.

A.4 Multifunctional software

d Chip

www.dchip.org/
The DNA-Chip Analyzer (dChip) package is for model-based analysis of oligonucleotide microarrays and can be used directly on data generated on the Affymetrix platform. The model operates on spot-level data, and provides tools for outlier detection, identifying and handling cross-hybridising spots and contaminating areas on the array, clustering and visualisation. The algorithm is described in Li and Wong (2001) and Schadt *et al.* (2001). Applications of d Chip are described in Hakak *et al.* (2001).

Expression Profiler

ep.ebi.ac.uk
Expression Profiler is a set of tools for clustering, analysis and visualisation of

gene expression and other genomic data. The user can perform cluster analysis, pattern discovery and visualisation, search based on Gene Ontology categories, extract putative regulatory sequences, study protein interactions and link analysis results to external tools and databases. EPCLUST, the clustering module of Expression Profiler, allows users to perform data selection, normalisation, randomisation, clustering, similarity searches and other operations with expression data. All tools in EP are accessible over the Internet via a Web interface. Calculations are performed on the server side. Use of these tools for automatic discovery of potential regulatory signals in genomes has been described (Vilo *et al.*, 2000).

J-Express

www.molmine.com/
J-Express is a portable software package for the analysis of microarray data. It accepts as input gene expression matrices or spot quantitation files. The program contains routines for normalisation and filtering to transform spot quantitation into log ratio gene expression matrices. Expression profiles can be explored using (two-ways) hierarchical clustering, *K*-means, clustering, self-organising maps, principal component analysis, and profile similarity searches. The program also contains a project management system complete with metadata for all derived datasets to allow for documentation of results (Dysvik and Jonassen, 2001).

MAExplorer

www.lecb.ncifcrf.gov/mae/
MAExplorer permits analysis of cDNA microarrays from mouse mammary tissue and databases from the Mammary Genome Anatomy Project (MGAP). The user is able to analyse expression of individual genes, gene families and clusters, and compare expression patterns generated from data in the linked databases.

maxdView

bioinf.man.ac.uk/microarray/maxd/maxdView/
maxdView is a modular analysis and visualisation environment for integrating existing analysis and display tools and for facilitating the development of new tools. maxdView can be run as part of the ISYS environment (ISYS is a flexible platform for the integration of bioinformatics software tools and databases: www.ncgr.org/isys/).

MeV (Microarray experiment Viewer)

www.tigr.org/softlab/
MeV is one of a number of bioinformatics tools developed by The Institute for

Genome Research (Rockville, MD). Gene expression data can be processed using a number of methods, including filtering, sorting, log transformation and normalisation, and the resulting data grouped using hierarchical clustering, self-organising maps, K-means, principal component analysis or support vector machines. A useful display tool provides an overview of the steps employed in each analysis.

TIGR Microarray Data Analysis System (MIDAS)

www.tigr.org/softlab

TIGR MIDAS is a data quality fitting and normalisation tool. Raw data can be analysed using multiple normalisation, fitting and transformations algorithms including Lowes (Locfit) normalisation, consistency checking and intensity-dependent z-score fitting (slice analysis).

References

Biedl, T., Brejova, B., Demaine, E. D., Hamel, A. M., and Vinar, T. (2001). *Optimal arrangement of leaves in the tree representing hierarchical clustering of gene expression data*. Technical report CS-2001-14 (Department of Computer Science, University of Waterloo).

Colantuoni, C., Zeger, S., and Pevsner, J. (2002). SNOMAD (Standardization and Normalization of Microarray Data): web accessible gene expression data analysis. *Bioinformatics*, in press.

Dopazo, J., and Carazo, J. M. J. (1997). Phylogenetic reconstruction using an unsupervised growing neural network that adopts the topology of a phylogenetic tree. *Journal of Molecular Evolution* **44**, 226–233.

Dudoit, S., Yang, Y. H., Callow, M. J., and Speed, T. (2000). Statistical methods for identifying differentially expressed genes in replicated cDNA microarray experiments. Available at
http://www.stat.berkeley.edu/users/terry/zarray/TechReport/578.pdf

Dysvik, B., and Jonassen, I. (2001). J-Express: exploring gene expression data using Java. *Bioinformatics* **17**, 369–370.

Eisen, M. B., Spellman, P., Botstein, D., and Brown, P. O. (1998). Cluster analysis and display of genome-wide expression patterns. *Proceedings of the National Academy of Sciences of the United States of America* **95**, 14863–14868.

Golub, T.R., Slonim, D., Tameyo, P., Huard, C., Gaasinbeck, M., Mesirov, J.P., Coller, H., Loh, M.L., Downing, J.R., Caligiuri, M.A. *et al.* (1999). Molecular classification of cancer: class discovery and class prediction by gene expression monitoring. *Science* **286**, 531–537.

Grant, G. R., Manduchi, E., and Stoeckert, C. J. J. (2002). Using non-parametric methods in the context of multiple testing to identify differentially expressed genes. In *Methods of Microarray Data Analysis*, S. M. Lin and K. F. Johnson, eds., pp. 37–55 (Kluwer Academic Publishers, Boston).

Hakak, Y., Walker, J. R., Li, C., Wong, H. W., Davis, K. L., Buxbaum, J. D., Haroutunian, V., and Fienberg, A. A. (2001). Genome-wide expression analysis reveals dysregulation of myelination-related genes in chronic schizophrenia.

Proceedings of the National Academy of Sciences of the United States of America **98**, 4746–4751.

Herrero, J., Valencia, A., and Dopazo, J. (2001). A hierarchical unsupervised growing neural network for clustering gene expression patterns. *Bioinformatics* **17**, 126–136.

Ideker, T., Thorsson, V., Siegel, A. F., and Hood, L. (2000). Testing for differentially-expressed genes by maximum-likelihood analysis of microarray data. *Journal of Computational Biology* **7**, 805–817.

Lazzaroni, L., and Owen, A. (2002). Plaid models for gene expression data. *Statistica Sinica* **12**, 61–86.

Li, C., and Wong, W. H. (2001). Model-based analysis of oligonucleotide arrays: Expression index computation and outlier detection. *Proceedings of the National Academy of Sciences of the United States of America* **98**, 31–36.

Manduchi, E., Grant, G. R., McKenzie, S. E., Overton, G. C., Surrey, S., and Stoeckert, C. J. J. (2000). Generation of patterns from gene expression data by assigning confidence to differentially expressed genes. *Bioinformatics* **16**, 685–698.

Newton, M. N., Kendziorski, C. M., Richmond, C. S., Blattner, F. R., and Tsui, K. W. (2001). On differential variability of expression ratios: improving statistical inference about gene expression changes from microarray data. *Journal of Computational Biology* **8**, 37–52.

Peterson, L. E. (2002). CLUSFAVOR 5.0: hierarchical cluster and principal-component analysis of microarray-based transcriptional profiles. *Genome Biology* **3**, 0002.1–0002.8.

Schadt, E. E., Li, C., Ellis, B., and Wong, W. H. (2001). Feature extraction and normalization algorithms for high-density oligonucleotide gene expression array data. *Journal of Cell Biochemistry (Supplement)* **37**, 120–125.

Tamayo, P., Slonim, D., Mesirov, J., Zhu, Q., Kitareewan, S., Dmitrovsky, E., Lander, E. S., and Golub, T. R. (1999). Interpreting patterns of gene expression with self-organizing maps: methods and application to hematopoetic differentiation. *Proceedings of the National Academy of Sciences of the United States of America* **96**, 2907–2912.

Tusher, V.G., Tibshirani, R. and Chu, G. (2001). Significance analysis of microarrays applied to the ionizing radiation response. *Proceedings of the National Academy of Sciences of the United States of America* **98**, 5116–5121.

Vilo, J., Brazma, A., Jonassen, I., Robinson, A., and Ukkonen, E. (2000). Mining for putative regulatory elements in the yeast genome using gene expression data. Paper presented at *Eighth International Conference on Intelligent Systems for Molecular Biology* (American Association for Artificial Intelligence Press, Menlo Park, CA).

Wall, M. E., Dyck, P. A., and Brettin, T. S. (2001). SVDMAN – Singular value decomposition analysis of microarray data. *Bioinformatics* **17**, 566–568.

Yang, Y. H., Dudoit, S., Luu, P., and Speed, T. P. (2001). Normalization for cDNA microarray data. *Proceedings of SPIE, BIOS 2001, Microarrays: Optical Technologies and Informatics* **4266**, 141–152.

Zhang, K., and Zhao, H. (2000). Assessing reliability of gene clusters from gene expression data. *Functional and Integrative Genomics* **1**, 156–173.

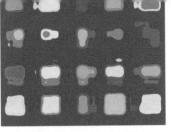

Glossary

The glossary describes terms used throughout the text and in the wider microarray literature and also seeks to explain some of the concepts and terminology proposed by the Microarray Gene Expression Data Society (MGED) as part of the 'Minimum Information About a Microarray Experiment' (MIAME) and the Microarray Gene Expression Markup Language (MAGE) (Brazma *et al.*, 2001). This terminology is more technical than that used on a daily basis within laboratories, but will need to be more widely understood if the proposed guidelines for the reporting and description of microarray data are adopted. Note that as with any emerging subject area, there is still little standardisation in the ways terms are defined, and so the following definitions may differ from those used elsewhere.

ALTERNATIVE SPLICING

Splicing is the process by which different regions of RNA are removed from a transcript and the remaining RNA is joined together before translation. See also http://www.mged.org/micml. If different regions, or combinations of regions, can be removed from the same mRNA, the transcript is said to be alternatively spliced. Alternative splicing of transcripts is one of the mechanisms by which a large number of proteins may be encoded by a smaller number of genes.

ARRAYEXPRESS

ArrayExpress is a public repository for gene expression data, based at the European Bioinformatics Institute (EBI).

BACKGROUND/LOCAL BACKGROUND

The signal intensity detected in regions of the array that do not correspond to features (spots) is referred to as the 'background signal'. Data processing usually includes a step in which the signal intensity value for each feature is corrected to account for the background, to obtain the 'corrected spot intensity' (CSI). This correction may be derived from a local calculation of the background, e.g. the background in the vicinity of the feature, and/or from the background across the whole of the array.

BIOSOURCE

Biological information relating to the source of the labelled extract. This includes the genus and species of the organism of origin, and other information that usually depends on the type of organism, e.g. strain, strain background, sex, age, organ, tissue, etc.

CIBEX (CENTRE FOR INFORMATION BIOLOGY EXPERIMENTATION DATABASE)

The public repository for gene expression data at the DNA Data Bank of Japan.

CO-EXPRESSION

Transcripts with similar expression patterns are said to be co-expressed. Co-expression may, or may not, indicate that the transcripts are co-regulated.

COMPOSITE SEQUENCE

A set of reporters that provide information on the expression of a single transcript or gene. An example is the set of 'match oligonucleotides' on an Affymetrix GeneChip^{TM} that hybridise to different parts of a single transcript and are used to derive a value that represents the relative expression level. Composite sequences are sometimes referred to as composite reporters.

CO-REGULATION

Transcripts with similar expression patterns that can be attributed to common regulatory mechanisms are said to be co-regulated. A typical example would be two genes that are activated by a common transcription factor.

DATA TRANSFORMATION

An operation applied to either raw or transformed data.

DYE-FLIP OR DYE-SWAP EXPERIMENTS

Experiments in which two or more hybridisations of labelled extract are carried out, such that each of the labelled extracts is labelled with each of the dyes, e.g.:
• Hybridisation 1, extract A is labelled with Cy5 and extract B is labelled with Cy3.

• Hybridisation 2, extract B is labelled with Cy5 and extract A is labelled with Cy3.

Data from dye-swap experiments may be thought of as replicate data. It is important to include dye swaps in the experimental design in situations where more than one labelled extract is hybridised to each array, and for determining which genes are differentially expressed between the extracts under study because of differential dye incorporation. Loop designs, which are similar in principle to dye-swap experiments, are another method of addressing this problem.

ELEMENT

Another word used in place of 'feature' or 'spot', i.e. the location and nature of the reporter on the array, where the reporter is the material that makes up the feature (e.g. the oligonucleotide or PCR generated sequence). Other terms used are 'target' and 'probe', although these are used differently in connection with different types of arrays and are thus ambiguous.

ERROR

Errors associated with measurement may be divided into two types:

1 *Systematic error.* Systematic error is also known as bias, and results in the consistent over- or underestimation of the true value. Sources of bias are factors such as differential incorporation of label or the location of a feature (spot) on the array. Normalisation is used to account for systematic error between data obtained from individual features and between labelled extracts.

2 *Random error.* Random errors reflect inevitable uncertainty in the measurement of data. Random errors cannot be removed, but may be reduced by reducing the number of variables in the experimental design, e.g. by processing samples at the same time of day, and keeping the strain background consistent across experiments. Estimates of the random error obtained from replicate datasets may be used to assign a confidence score for assessing the statistical validity of a measurement.

ESTS (EXPRESSED SEQUENCE TAGS)

Short unique DNA sequences (200 to 500 base pairs), 'expressed' as mRNA and whose location and sequence are known.

EXON

An exon is a region of a gene that is used as a template for translation, i.e. a region of a gene that encodes for protein.

FACS (FLUORESCENCE ACTIVATED CELL SORTING)

A technique for separating cell populations in which some cells fluoresce and others do not. The fluorescence usually derives from expression of a reporter construct, or from labelling of a subpopulation of cells.

FEATURE (SPOT)

The location and nature of the reporter on the array, where the reporter is the material that makes up the feature (e.g. the oligonucleotide or PCR generated sequence). Other terms used are 'target' and 'probe', although these are used differently in connection with different types of arrays and are thus ambiguous.

FEATURE EXTRACTION

The process in which the information describing the feature (spot) is extracted from the scanned image.

FILTER ARRAY

Arrays or chips may be printed on a number of different surfaces, including nitrocellulose. These are sometimes referred to as 'filter arrays' to distinguish them from those on glass, or silicon, substrates.

FILTERING

Filtering is a way of identifying subsets of data that fulfil defined criteria. It can be carried out in numerous ways to identify the data of interest. Filtering is most commonly used to remove values that fall below or above a certain threshold. An alternative to filtering is to 'floor' the data (see below).

FLAG (OR 'TAG')

Data may be flagged, or tagged, to remind the data analyst about a property of the data that may be important for subsequent interpretation of the results. Flagging has the advantage that the particular piece of data is not removed from the dataset, but bears a 'mark' to distinguish it from the rest of the data. Examples include flagging of 'low confidence' data, outlier data, data that have been floored, and imputed data.

FLOORING

The process by which scaled data are adjusted so that values that fall below a certain number of intensity units are brought up to the value of the 'floor'. An example is when fluorescence intensities of below 50 are raised to 50. This is a way of including data for genes that are expressed at a low level without using the actual value obtained, which could be largely due to noise. Flooring is often carried out before the fold change in gene expression is calculated. Flooring produces apparently smaller fold changes for genes expressed at a low level.

The parallel process by which genes whose expression level is high are brought down to a value is sometimes called setting a 'ceiling'. This permits inclusion of data from genes whose expression level is high but probably cannot be calculated accurately, e.g. if the signal is likely to be saturated.

FUNCTIONAL GENOMICS

The subject areas known collectively as functional genomics consider biological systems on a genome-wide scale and include analysis of genome sequence, gene prediction, identification of gene function, genes and gene product interactions, etc., as well as comparisons across genomes.

GENECHIP^{TM}

A trademark of Affymetrix used to describe a type of oligonucleotide-based array that is synthesised *in situ* on a silicon substrate by photolithography.

GEO (GENE EXPRESSION OMNIBUS)

GEO is a gene expression and hybridisation array data repository, as well as an online resource for the retrieval of gene expression data from any organism or artificial source. GEO is run by the National Center for Biotechnology Information (NCBI). Further information can be obtained at http://www.ncbi.nih.gov/geo/.

INTRON

A non-coding region of a gene, i.e. a region that may be removed from the corresponding transcript before translation.

LABELLED EXTRACT

A term used to refer to the labelled population of nucleic acid that is hybridised to the array. Terms such as 'target' and 'probe' should be avoided as they are used to refer to both the labelled nucleic acid in solution and the nucleic acid attached to the array (feature/spot), depending on the microarray platform.

LIMS (LABORATORY INFORMATION MANAGEMENT SYSTEM)

A database/data warehouse for recording technical aspects of array construction, design within the laboratory and other experimental information. LIMSs are frequently used to keep track of which nucleic acid is present on each array, where it is stored in the laboratory, the source of the nucleic acid, etc.

MAGE (MICROARRAY GENE EXPRESSION)

The 'Microarray Gene Expression' working group (http://www.mged.org/Workgroups/MAGE/mage.html) is a group working towards setting standards for the representation of microarray expression data via the establishment of a data exchange model (MAGE-OM: Microarray Gene Expression – Object Model) and data exchange format (MAGE-ML: Microarray Gene Expression – Markup Language). MAGEstk (MAGE Software Toolkit) is a collection of packages that act as converters between MAGE-OM and MAGE-ML.

MEDIAN

The value in a set of data such that half of the values are above it and half are below it. The median fluorescence intensity of each chip is often used to derive a scaling factor for between-array comparison.

MIAME (MINIMUM INFORMATION ABOUT A MICROARRAY EXPERIMENT)

MIAME is an informal specification that has been put forward by members of the Microarray Gene Expression Data Society (see http://www.mged.org/miame) to guide cooperative data collection and description (Brazma *et al.*, 2001). The object of MIAME is to ensure optimum interpretability of experimental results, and to facilitate independent verification and data exchange and the establishment of databases and public repositories of data.

MIAME can be thought of as (i) an attempt to formulate the minimum information that should be included in a description of a microarray experiment to enable unambiguous interpretation, or potential verification, of its results; and (ii) a description of the information that would be required by an experimenter from a different laboratory for repeating an experimental method and comparing the results. MIAME is not a formal specification and items that make up the 'minimum information' will be different for different datasets. The terms used in this book follow conventions endorsed by MIAME, wherever possible.

MIAME COMPLIANT

Microarray data that include information, or references/links to information, required by the MIAME document are said to be MIAME compliant.

MIAMEXPRESS

A Web-based submission tool created at the European Bioinformatics Institute (EBI) for submitting data to ArrayExpress. The tool supports the generation of MIAME supportive data in XML. Further information may be obtained at http://www.ebi.ac.uk/microarray/.

MICROARRAY GENE EXPRESSION DATA SOCIETY (MGED)

The MGED is an international organisation for facilitating the sharing of microarray data from functional genomics and proteomics experiments. This includes setting standards for DNA array experiment annotation and data representation, and introduction of standard experimental controls and data normalisation methods. This will facilitate the establishment of gene expression data repositories, enhance the comparability of gene expression data from different sources and the inter-operability of different gene expression databases and data analysis software. Further information can be found at http://www.ebi.ac.uk/microarray/MGED/.

OLIGONUCLEOTIDE

A short sequence of DNA (usually 80 or fewer nucleotides). Within the context of a microarray, this sequence is used to provide information on a transcript of complementary sequence, via a hybridisation reaction. Features (spots) on arrays are usually made up of oligonucleotide sequences, or longer DNA sequences generated by PCR, representing transcripts or fragments of transcripts.

ORF (OPEN READING FRAME)

An open reading frame is a region of sequence, interrupted by a stop codon, that encodes all or part of a protein.

OUTLIER

An extreme value in a distribution. Outliers may represent informative or artefactual data.

POLY A RNA

In eukaryotes most messenger RNA (the RNA that encodes protein) ends in a string of riboadenosine triphosphates. This characteristic of mRNA distinguishes it from ribosomal RNA (rRNA) and transfer RNA (tRNA) and is a property commonly exploited in order to isolate mRNA from total RNA. Bacterial mRNAs do not have poly A tails, so mRNA isolation from bacterial samples is carried out using other methods. An artificial poly A tail is added to the bacterial control mRNA that is sometimes 'spiked in' to total RNA from eukaryotic cells during sample preparation. This ensures that the bacterial RNA is represented in the sample and that non-modified contaminating bacterial RNA does not contribute to the signal from the control spots.

PROBE

In the context of spotted arrays the term probe often refers to the labelled population of nucleic acid in solution, while in connection with GeneChips™ it is used to refer to the nucleic acid attached to the array. To avoid confusion, here we use the MIAME convention in referring to the mobile population of nucleic acid as the labelled extract and the nucleic acid attached to the array as the reporter, feature or spot.

REFERENCE ARRAY/CHIP

Data from a 'reference' or 'baseline' array may be used as the basis for comparing data obtained from hybridisation of labelled extracts to other arrays. A typical way to use the data from a 'reference chip' is to scale, or normalise, all the data to that of the reference, based on a property such as global intensity.

REFERENCE SAMPLE

A reference sample is the sample that others are compared with. Reference samples are frequently used with spotted arrays where two or more labelled extracts are hybridised to a single array. A commonly used reference sample is genomic DNA.

REFERENCE SET

Arrays whose data have been scaled together can be compared directly and form a reference set.

REPLICATE 'SPOTS'

There are two types of replicate features – those for which the element (the sequence represented) is the same and those for which the element is different. Where the reporter (the nucleic acid molecules that make up the feature) is different, the replicates may be referred to as a composite sequence or composite element.

REPORTER SEQUENCE

The MIAME/MAGE convention uses 'reporter' or 'reporter sequence' to refer to the set of molecules that make up a feature or spot. This may be a set of oligonucleotide or PCR generated sequences. Note that the same reporter can be at multiple locations on an array, while a 'feature/spot' cannot, because the latter term includes information about the location of the nucleic acid on the array.

SAMPLE

The sample is the biological material from which the gene expression profile was obtained. A description of a sample would typically include the source of the material (e.g. genus, species, the tissue or cell type, also known as the 'biosource') and the conditions or treatment applied. Nucleic acid is extracted from the sample and labelled to obtain one or more labelled extracts that are hybridised to the array. Note that a distinction is sometimes made between a primary and a derived sample: a derived sample may consist of multiple primary samples, for example where samples are pooled.

SATURATION

The measured signal intensity, usually representative of the gene expression level, increases in proportion to the increasing concentration of the labelled molecule (nucleic acid in the labelled extract) over the 'linear range', and at some point reaches saturation. At this point the measured expression level of the gene does not increase further.

SNP (SINGLE-NUCLEOTIDE POLYMORPHISM)

Single-nucleotide polymorphisms are positions in a genome at which alternative bases occur at significant frequency. SNPs are the most frequent type of sequence variation found in the human genome, and knowledge of the location and frequency of SNPs is therefore exploited extensively in medical genetics.

SPIKED-IN CONTROLS

Exogenous controls that are added at a stage in the preparation of the labelled extract and are designed to hybridise to complementary features on the array. The information obtained from spiked-in controls may be used for normalisation between labelled extracts, for assessing the sensitivity with which gene expression can be detected, and/or as a calibration reference to obtain absolute expression levels and extend the linear signal intensity range for scanning.

SPLICING

The process by which pieces of mRNA are cut, a segment is excised and the remaining sequence is joined together. Splicing permits greater protein diversity from a discrete amount of DNA, as mRNA can be cut and rejoined in different combinations. Differential splicing contributes to the observed physiological complexity of cells.

SPOT (FEATURE)

Spot is used as a generic term to refer to the nucleic acid that is attached to the array. MAGE uses the word 'spot' (or feature) to describe the location and nature of the reporter on the array, where the reporter is the material that makes up the feature (e.g. the oligonucleotide or PCR generated sequence). Other terms used are 'target' and 'probe', although these are used differently in connection with different types of arrays and are thus ambiguous.

TARGET

In relation to spotted arrays the term target is used to refer to the nucleic acid attached to the array and the word 'probe' to refer to the labelled population of nucleic acid with which it is hybridised. The terms usually have the reverse meaning when referring to GeneChips^{TM}. We have tried to avoid use of the term target and instead refer to the labelled, mobile population of nucleic acid as the labelled extract and the nucleic acid attached to the array as a spot, feature or reporter.

TARGETED CELL TYPE

The target cell type is the cell type of primary interest. The biomaterial may be derived from a mixed population of cells allthough only one cell type is of interest.

TILING ARRAYS

Arrays in which overlapping sequences are represented by the features (spots) on the array. These are commonly used to identify the 5′ and 3′ ends of transcripts, for gene discovery and annotation.

TRANSCRIPTION FACTOR

A factor that regulates transcription (the making of an RNA copy of a DNA segment). Usually a protein, or multi-subunit complex.

TREATMENT OR CONDITION

The condition or treatment describes 'what has happened' to the biosource, i.e. the biological material under study, before preparation of the labelled extract. A sample may be described in terms of the biosource and the treatment.

VENN DIAGRAM

A graphical method of representing data that have been assigned to different classes. A 'flat' method of clustering.

Reference

Brazma, A., Hingamp, P., Quackenbush, P., Sherlock, G., Spellman, P., Stoeckert, C., Aach, J., Ansorge, W., Ball, C. A., Causton, H. C., *et al.* (2001). Minimum Information About a Microarray Experiment (MIAME) – toward standards for microarray data. *Nature Genetics* 29, 365–371.

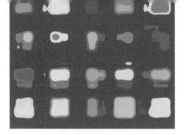

Index

Page numbers in *italics* indicate figures and those in **bold** indicate tables, where these are separated from their main reference in the text.